BY RODNEY BARKER

AND THE WATERS TURNED TO BLOOD

DANCING WITH THE DEVIL

THE BROKEN CIRCLE

THE HIROSHIMA MAIDENS

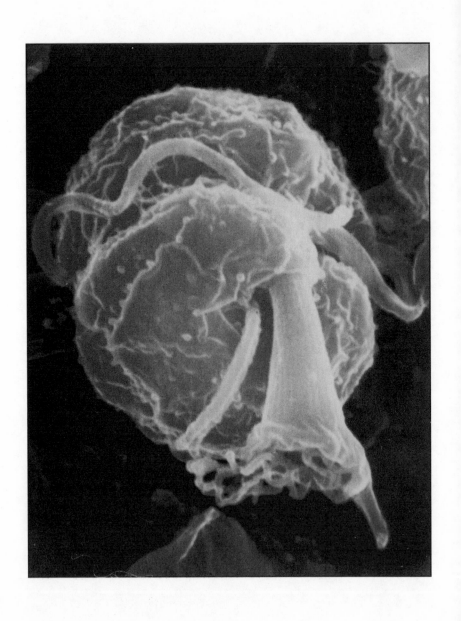

AND THE WATERS
TURNED TO
BLOOD

The Ultimate Biological Threat

RODNEY BARKER

*For George Watson —
Who told this story made you angry —
it did me too. Thanks for your interest!*

Rod Barker

SIMON & SCHUSTER

SIMON & SCHUSTER

Rockefeller Center
1230 Avenue of the Americas
New York, NY 10020

Designed by Jeanette Olender
Manufactured in the United States of America

10   9   8   7   6   5   4   3   2   1

Photo facing title page: The killer-dinoflagellate, *Pfiesteria piscicida*,
in its most toxic stage. Photo by Howard Glasgow.

Library of Congress Cataloging-in-Publication Data
Barker, Rodney.
And the waters turned to blood : the ultimate biological threat /
Rodney Barker.
p. cm.
1. Dinoflagellate blooms—Health aspects.   I. Title.
RA1242.M34B37   1997
615.9′52987—dc21   97-86   CIP
ISBN 0-684-83126-0

To my friend and editor, Michael Korda

Contents

AND THE WATERS TURNED TO BLOOD

F O R E W O R D

"What's wrong with me, Mommy?"

It was shortly after the Jorgensen family returned from their 1995 North Carolina vacation that Gail Jorgensen, a thirty-nine-year-old mother of two, began to suffer from a mysterious malaise. "I had no idea your body could even produce such bizarre symptoms," she would say later of the chills that started at her waist and seemed to turn the blood in the lower half of her body cold, the tingling in her hands and feet that could have been someone pinching her, the muscle spasms that ignited an intense burning sensation.

They seemed to run a course: She would be perfectly normal one day, the symptoms would start to appear the next, and they would build to an excruciating peak, only to die down and almost go away before beginning again.

At times it seemed that there was something alien alive inside her, fighting for control.

She was reluctant at first to talk much about these sensations, because she was afraid that unless others had experienced something similar, "they would think I was a nutcase," and because her two children began to mimic her reported symptoms. Her son, Soren, aged six, said he was having trouble sleeping because his legs were shaking. Her twelve-year-old daughter, Carolyn, spoke about how, after she went for a walk, her arms and legs felt as if they were on fire.

When she heard her children talk about their symptoms, initially Gail figured they were just developing sympathy pains. But when they continued to fuss after she silenced herself, she listened more closely.

And she was struck by the fact that when they described what they were feeling, not only could she relate to what they were saying, but their language was more vivid and accurate than her own. Soren's "It feels like electricity is going through my legs, Mommy," exactly expressed her own sensations.

While the uncanny rightness of her children's vocabulary suggested they were not making things up, Gail Jorgensen wanted proof. "Honey, the next time you have muscle spasms, I want you to show them to me," she told her daughter.

"Okay," Carolyn replied.

Later that day she heard her daughter call from her room: Come quick, it was happening again. And sure enough, the muscles in Carolyn's legs were going through contractions—so severe it looked as though a small animal were scurrying around under her skin.

Proof that Soren wasn't inventing symptoms came the morning he crawled into bed with his parents because he was having difficulty sleeping. In fact, Gail was having her own problems with insomnia during this period. Just as she was dozing off, her body would start twitching, and the only thing that would relax her was a hot bath. Sometimes she would take as many as three baths a night. On this particular morning, no sooner had Soren drifted off to sleep beside her than his arm flew up, then his body convulsed as though jolted by a shock.

Now that she knew her children's disorder was similar to hers, Gail Jorgensen became obsessed with finding the cause. When they asked her, "What's wrong with me, Mommy? Why is this happening?" she wanted to be able to give them a better answer than "I don't know."

She turned to the medical community for help, and over the following months she was relentless in her pursuit of an explanation. After listening to her try to put her health problems into words, her primary care physician disqualified himself, referring her to an internist, who passed her to a neurologist, both of whom were equally puzzled. One said it sounded like a central nervous system disorder, the other hypothesized that she must have been exposed to a toxin, but neither could say for sure. Both did tests, but a brain scan and an EEG came up negative, and her blood work was fine.

The closest she came to thinking she had it solved was when she had a conversation with a friend who had suffered from multiple sclerosis for fifteen years; in the beginning of her disease, she too had experienced tingling and burning, and no one could figure out what was going on until finally she was checked for MS. Just the thought of MS, especially with two children to care for, terrified Gail; but if that was the source of her problems, she wanted to know so she could make plans. And if it wasn't, the peace of mind was just as important. So she went through another battery of tests, which, to her relief, eliminated MS as a possibility. The flip side was that it added to her frustration: she was no closer to an answer.

There followed a succession of dead ends. One of the doctors who examined her suggested that maybe the source of the problem was environmental, so she paid six hundred dollars to have the air quality in her house tested. Nothing wrong there. Then, because their home was a veritable menagerie—they had birds, rabbits, and a dog—she checked with a veterinarian to see if there were any animal-borne diseases going around. None that he was aware of. Next she called the Centers for Disease Control in Atlanta, to inquire about the possibility that they were infected with parasites. The symptoms didn't match up. Someone recommended she see a food allergist, thinking perhaps their diet was at fault, and she did, despite the fact that her husband ate at the same dinner table and wasn't on the sick list.

After months of subjecting herself to every examination imaginable, with nothing to show for it—to spare her children more discomfort, she had the doctors run their tests on her—she consulted a pediatrician. "I don't know what's going on," she said, "but we're all sick here, and I have to know what we're dealing with. Can you please help us?"

He talked to her for some time, then shook his head; as far as he was concerned, it was a medical enigma. But if *his* kids were in a bad way, he said, "I know what I'd do. I'd go to the medical center at Johns Hopkins for a checkup. They're the best in the country."

Gail Jorgensen took that advice, and a neurologist at John Hopkins ran a series of tests. All of them were inconclusive, and during the follow-up interview, the neurologist made a recommendation. "Please don't take this wrong, but maybe what you really need to help you through this is to see a psychotherapist."

This was not the first time that someone had suggested to Gail that her problems might be mental. In fact, by this time she had become pretty much isolated from her family. While her husband had continued to be a sympathetic supporter, her sister, a registered nurse, had become convinced that she was simply stressed out: what she really needed was a shrink, and a little medication wouldn't hurt.

"I am not stressed out, I do not need a psychotherapist, and I don't want any pills," Gail let everyone know in as even a voice as she could maintain. "I know my body better than any doctor, and I'm telling you my body is telling me that something is wrong here. And my children's bodies are telling me there is something wrong here."

It was all so vexing, and she felt so helpless, and in the middle of this ordeal Gail Jorgensen turned forty. She would remember it as the worst year of her life. Month after month of constantly fighting for understanding and getting nowhere had brought her to a point, on several occasions, where she had said to herself, *I cannot go on living like this.*

Emotionally exhausted and fearful of a nervous breakdown, she sat down one day with her children and said, "Okay, kids, we're going to play a game. It's called Backtrack. Let's put our heads together and try to remember when this all started and where we were just before we began to feel all these weird things."

It was as easy as unfolding a road map. Where they had been was North Carolina, and what they had done was a lot of sightseeing. Just before the onset of the symptoms, the family had visited Raleigh and Rocky Mount and points of interest around the state. Gail was a nature buff and an animal lover—as a child, she would catch frogs and walk right up to snakes in the woods as if they were her pets—so she was drawn to the state parks more than to the cities. That was why, at one stop, she had walked her kids down to the edge of one of North Carolina's estuarine rivers to look for snails. The water had been so low, she remembered, that they were able to hopscotch their way on the backs of rocks almost to the middle of the river, where they had spent several hours exploring and playing in the water.

Now Carolyn said, "Mom, what about all those dead fish?"

Out on the river, they had taken their shoes and socks off and rolled up their pants; they were wading in the shallows when Gail noticed that the river was full of floating fish. They looked like autumn leaves scattered on a lawn after a big wind. She told her children not to touch them, but there were so many it was hard to avoid them completely. Gail Jorgensen had forgotten this, but now she exclaimed, "Oh my God, that's right!"

In the aftermath of her excitement over the possibility that there was a connection between the dead fish and their problems, Gail Jorgensen found herself remembering the only other time she had seen dead fish in massive numbers. It had been in Florida, when she was young, on a vacation that was spoiled by a red tide. Now she rushed to the local library to look up red tides. She learned that they were brought on by a microscopic form of life that emitted a poison deadly to fish. But according to the reference material, red tides occurred in the ocean, not in rivers, and the effects on humans were primarily respiratory, which had not been her experience. Maybe the dead fish were not the missing link.

Where should she turn next? The North Carolina Department of Health seemed a logical source of information. It took several calls before she was put in touch with the right official, and she asked him if anybody else had reported similar symptoms from contact with dead fish on North Carolina's rivers. The official was polite but adamant: no such phenomena had been reported. She asked if he would mind double-checking to be sure, because this was the only thing that made sense to her. After giving her address and phone number, she asked out of curiosity what was killing the fish. She was told it must have been because the water was low. Now, that didn't make sense, so she asked if the waters in which fish were dying in such large numbers had been tested for toxins. Those studies were being conducted, he replied, and she asked him to please notify her of the results.

No one ever called. And even though Gail continued to phone state health officials, asking for updates, none were given.

Gail Jorgensen felt intuitively that something in that river—most likely the same thing that had killed the fish—lay behind her family's health problems. A small but telling piece of evidence was that her husband had been with them that day but had watched them from the riverbank. It was the only time she could think of during the trip that they had not done something as an entire family.

"Are we going to end up dead like the fish?" her daughter asked.

"I don't think so, baby," Gail replied. "Fish are a lot smaller than we are, and our bodies have a way of fighting things off. So no, I don't think we have to worry that what killed them will hurt us."

But she wasn't absolutely sure.

It was at the six-month mark that she realized she was starting to feel better. It seemed that her children were also complaining less and less, and she began to think, *Thank goodness. Whatever it is, it's working its way out.*

But then she noticed something strange going on with her son. They would be sitting in a room together, talking, and Soren would drift off. To almost anyone else it would have looked as if he were merely daydreaming, but his mother knew this was something different. When she reached out and tapped his face and said, "Soren? Soren?" she couldn't get his attention. Ten or twenty seconds later he would snap out of it, and he remembered nothing that had gone on during the elapsed time.

Something's not right here, Gail thought, and she started monitoring him closely. Several nights later, she was awakened from sleep by footsteps upstairs. Rising, she traced the noises to Soren's room, only to find him asleep in bed. Wide awake now, she decided to stay and watch him awhile, and lo and behold, less than an hour later he sat up, got out of bed, and began to walk around the room. "Soren?" she called softly, but he did not respond. His eyes were open and his body was moving, but it was as if his brain weren't there.

She drove him to the hospital emergency room, where it happened again. The attending doctors identified it as a petit mal seizure, a diagnosis that was supported by tests. But what the doctors were unable to figure out was the cause. All she got from them was: "This is something weird. I've never seen anything like this before."

It was the same line she had heard from doctors throughout her search for an understanding of her own bizarre symptoms.

A year and a half has passed since this nightmare began for the Jorgensen family, and Gail lives with the fear that it is not over yet. The Jorgensens are of hardy Scandinavian stock and used to enjoy good health, but now they have a hard time fighting things off. It seems they catch every contagious illness that goes around—colds, flu, you name it. To this day no one has been able to tell Gail what the trouble is, but about where the trouble had its origin she has no doubts. She is convinced that there is something dangerous lurking in the waters of North Carolina. She believes there are people in the state who are aware of it. And she suspects that they know more about the threat it poses to human health than they have been willing to admit publicly.

And she is right.

PART ONE

"I've been waiting to meet you

for a long time."

1

The circumstances that brought Dr. JoAnn Burkholder to North Carolina in 1986 were so sudden and surprising that it almost seemed as if the Tar Heel State dialed her number, rather than the other way around.

She was thirty-three and close to completing her doctorate in botanical limnology—the scientific study of plants that thrive in freshwater lakes and streams—at Michigan State, and for months she had devoted her Saturdays to filling out job applications. The effort had produced several offers, but none she could bring herself to accept. A faculty position at Fordham University came with a low salary, and the thought of living in New York City made her writhe as if from physical confinement. Michigan State had asked her to apply for a postdoctoral fellowship, and it would have been a comfortable position—a house in the woods and a year of writing papers—but she knew part of the reason it had been offered to her was that her great-grandmother was a full-blooded Cherokee. While it would have helped the administration to satisfy its minority quota, she didn't feel right about accepting it under those conditions. Not only would she be taking away an opportunity meant for people who were struggling, it was also important to her that wherever she went, her appointment be based on her merits as a scientist.

Then a professor from her undergraduate days at Iowa State University phoned, saying he had received a call from the head of a faculty search committee at North Carolina State University: they were having

difficulty attracting qualified applicants for an aquatic botanist position. It was May, she had been sending out applications for months. In
her field most job prospects came through in January or February, so
even though she said she would look into it, she thought it was too
late to apply. But then the professor called back to say he had recommended her for the position and they were personally requesting an
application. Hearing that, she remembers, she was so taken aback she
looked at the telephone receiver as if it were something with a will of
its own.

She had never been to North Carolina, and when she flew in for an
interview she had very few preconceptions. She didn't even know what
the landscape would look like. So she took a window seat on the flight
to Raleigh, and looking out the window, she noticed the thick pine
forests that began at the mountains and seemed to cover the northwest
part of the state like a wilderness. They made her smile, because she
loved the woods. Raised in Rockford, Illinois, she had been especially
close to her father, who had all but grown up in the woods, hunting,
fishing, and trapping during the Depression to help his family put
food on the table. When he married, he had badly wanted a son who
would share his interest in the outdoors, but there'd been no such luck.
His first daughter, Norma, would have been Norman, and John turned
out to be JoAnn. To his delight, however, so naturally did JoAnn take
to the open air that she almost made him forget she wasn't a boy. Her
happiest memories were of their hunting trips together, when he
taught her how to track animals and flush game, and would tease her
that she was his rabbit-and-squirrel retriever. Years later, she would
credit those early experiences in the woods as inaugurating her scientific curiosity about nature: at the age of nine, she was keeping notes
on how many bird species she could identify, and whenever she found
a nest with eggs, she would count and measure them, and record color
differences.

Also in view from the air were North Carolina's lakes, and they were
very different from what she was used to in Michigan. Not only were
most of them river impoundments, as opposed to natural bodies of
water, they also shone bright orange and brown—indicating a lot of
turbidity. Her specialty being the study of freshwater plants—where

and why and how they grow, with a particular emphasis on the way nutrients influence botanical communities—she knew just from looking that agricultural runoff and urban development were the likely culprits. She felt there was interesting work to be done here.

Not everyone can look back and point to a single experience or something read that gave his or her life direction and meaning, but JoAnn Burkholder could. When she was a junior in high school, she picked up a copy of *Life* magazine that featured a special report on "The Blighted Great Lakes: The Shocking Case of Our Inland Seas Dying from Man-Made Filth." Accompanying a text that recalled the deep, clear waters that inspired Longfellow to write "By the shore of Gitche Gumee" were sickening photographs of dead fish floating on currents that carried soapsuds, sewage, and chemicals, and of shorelines littered with abandoned cars and oil drums. While it was apparent from the article that the errors and negligence of we the people were responsible for fouling the lakes, it was beyond the scope of a sixteen-year-old conservation-minded girl to know how to approach the human aspect of the problem. But what did capture her interest was the concept that every lake had its own natural life cycle, beginning as a cold, clean body of water that supported a variety of healthy plant and animal life, and moving through stages toward becoming a shallow, warm marsh that hosted only the lowest forms of life. And while under normal conditions that process was supposed to take place over thousands of years, man-made pollution accomplished it in a fraction of the time by introducing nutrients that fed forms of plant life, particularly algae, that could, literally, choke a lake to death. After that article, there was never any question of what JoAnn Burkholder was going to do with her life. She was going to be involved in "water science" of some sort.

The people on the interview committee at North Carolina State were candid about their need: they were replacing a retiring professor and wanted someone forward-looking. Her former professor had told them that she filled the bill. She had a reputation for using advanced techniques in her research work because she was willing to go out on her own and learn what was necessary to accomplish a specific task. If she was involved in research that would benefit from electron microscopy, she took a course. If she needed to learn how to do autoradiography, she sought the people who could teach it to her—even if it meant going beyond the skill level of her professors.

When she asked the interviewers what would be expected of her, she found the position allowed for enormous flexibility. She could do almost anything she was interested in doing that pertained to the basic science of her field, they told her.

It was the first position she had interviewed for that she really wanted, and the feeling was mutual. Within three months, she was moving to Raleigh as an assistant professor of aquatic botany.

At that point it seemed as if fortune was working on JoAnne Burkholder's behalf. It didn't even faze her that in her haste to get settled she took a lease on a one-room flat in north Raleigh that she would come to think of as the Apartment from Hell. It was tiny and ugly, a single cell in an enormous honeycomb, with a prime view of rush-hour traffic. Making matters worse, on the other side of her wall lived someone who snored like a troll. She had to run a fan at high speed in order to get to sleep, and there were times when her apartment seemed to shake from the rumble of his snoring. She tried leaving notes, but they went unanswered, and one especially bad night she stormed into the hallway and pounded on his door. For ten minutes. It took her that long to realize he couldn't hear her over the din he was creating. By morning she was feeling homicidal. She'd taken the apartment because it had been the only place she could find near campus that did not allow pets and was within her budget. "I'm being held hostage to my allergy to cats," she would explain to those who asked her why she didn't just move.

Nevertheless it was home, and she succeeded in making it comfortable with pictures—an oil painting of a cabin in the woods, another of Brigadoon, and one from the tales of King Arthur. She was striving for a magical and timeless ambience, a place whose doorway was a moat that, once crossed, gave onto a fantasy world where she could forget everything unpleasant on the outside. It was an escape she needed, because as she was quickly discovering, an assistant professor at the university level was a lot like a pledge in a fraternity: a period of hazing preceded acceptance as an equal.

Although technically she was on a tenure track, she knew she had only five years to make good—which meant that she had to hurry up and write grant proposals and get them funded. The time had passed when tenure for newly hired professors was based on getting papers accepted in reputable scientific journals. Now you were expected to do that *and* go out and raise money to support your research. Plus set up a research laboratory and develop courses from scratch. On top of which you were also presented with, and had no choice but to accept, the junky jobs that were tossed your way by your department head— such as giving the guest lecture he'd been invited to deliver on botany appreciation at a community college a day's drive from Raleigh.

But JoAnn Burkholder was no stranger to adversity when it came to educational advancement. Indeed, she'd had to vault a lot of hurdles to get to this stage of her career.

Although she had been a precociously bright and quick-witted child, she was brought up in a working-class community where girls were discouraged from having ambitions beyond motherhood. In the second grade, after outscoring every other student on a test, she was told by her teacher that boys were endowed with native smarts girls did not possess, and as soon as their interests shifted from the playground to the classroom, they would surpass her. Her response had been, *No they won't!*

Even though she graduated from high school with straight A's, her mother had refused to sign college financial aid papers. As a result, JoAnn had worked as a drugstore clerk and scraped dishes in the school cafeteria to pay for her higher education.

When, as an undergraduate, she was hired as an assistant in a limnology research lab at Iowa State, she was the only female on staff. She was forced to put up with the good-old-boy chauvinism that came with an academic discipline traditionally considered the province of males because so much of what went on took place outdoors. But she showed that she could hike in the woods, ford streams, and camp with the best of them. In the end, it had been her limnology professor who recommended her to N.C. State.

For the first four months in North Carolina, she did not even have her own office. A space was cleared for a small desk in the corner of a storeroom. The conditions were so cramped that, at her own expense, she flew back to Michigan State to use the computer there in writing her grant proposals. And whereas in some institutions department members would take new people under their wing, offering them recommendations and leads, she was given very little mentoring—other than the understanding, in no uncertain terms, that she'd better get something funded fast.

All she did that first year was work. On a typical day, she arrived at the office at seven in the morning and was still there at midnight. She wrote proposals for stream and reservoir research, because she knew fresh waters best. She also went to the library and ferreted out the nearby professors in her field and their publications, so she could talk intelligently to them. She started making telephone calls: Hello, my name is JoAnn Burkholder. I'm new in the area. I understand you're a fishery ecologist. Your work sounds interesting—would you like to get together and talk about the aquatic ecology in the state? I'll buy lunch.

Other than making the rounds and visiting people related to her work—and leaving herself time to jog and to swim at the university gym, which she did religiously—she got out very little. She felt extremely isolated, but whenever she tried to think about activities she could engage in that would introduce her to new people, there was either a time conflict or a reason not to. She would have liked to join an organization, for example, but the only groups that interested her were environmentally oriented, and she feared affiliating herself with them might compromise her scientific integrity in the eyes of her colleagues.

As for a social life, it too was nonexistent. That had nothing to do with her ability to attract men. An athletic and shapely five feet five, with curly brown hair and electric-blue eyes, JoAnn had never been lacking in male admirers. It was more that at this point in her life she wasn't especially interested in getting involved with someone; in fact, her romantic history had left her slightly wary of the opposite sex.

There were men who became obsessed with her. David Mangold was the first name that came to mind, though it was a name she wished she could forget. She had been fifteen, he was two or three years older, and for some reason he fixated on her. For three and a half long years he pursued her, as what today would be called a stalker. He called her on the phone and wrote her letters, professing his love. And when she tried to tell him, reasonably, that she didn't share his feelings, he threatened to commit suicide, to kill her, to kill her parents. Once, reading a book in her room, she looked up and saw him staring at her through the window. An ACE radio operator, he would tap into her phone calls and interrupt angrily if the call was from another boy. When she came home from dates, he would be sitting on the front porch, waiting for her. Her parents tried everything they could to get rid of the boy, and his own family tried too. There was even an attempt to get a restraining order against him. But nothing worked until she moved to another state to go to college.

Still, years later, it wasn't fear of a relationship that prevented JoAnn Burkholder from going out socially, so much as priorities: she had a hard time relaxing and enjoying herself while her professional life was out of sorts. And to be completely honest, it seemed that whenever she tried to mix the two, she ended up miscalculating. A prime example was the suggestion she took from a professor. "Why don't you teach my aquatic ecology seminar. That way you'll get to meet new people, and gain experience in the process." She fell for it and taught the course. And while he was paid for the credit hours, she got a lesson learned.

And so it went for almost two years. She was awarded several small grants and even a couple of sizable ones, but it was still a struggle. Which goes a long way to explaining why she reacted as she did when she received a phone call from Dr. Edward Noga, a fish pathologist at the North Carolina State veterinary school, asking for her help in solving a mass murder.

2

The modern three-story building that is North Carolina State's College of Veterinary Medicine sprawls across a bucolic sweep of rolling green fields several miles west of the main campus. Its offices, classrooms, and research laboratories are approached by a red-brick walkway that leads from a parking lot to the front entrance. First-time visitors often stop on their way into the vet school to read from a row of bricks etched with names: donors to the college, and pets to whom contributions to the building fund were dedicated. *In Memory of Pepper—Don and Nancy. In Memory of Slugger and Ted— The Biggs Family.* Perhaps it is because they think of the inscribed bricks as little gravestones, but most people make a point of stepping over that row rather than on it.

In back of the main building are four towering grain silos and the laboratory-animal houses, which are referred to as finger barns because they are long and narrow and there are five of them. Horses and pigs and cattle, dogs and cats, are kept there, and the last one down— Building 311—is used for poultry and fish. This is where Dr. Edward Noga conducted studies into the causes of fish diseases and developed treatments, though it was actually his Ph.D. student Stephen Smith who was the first to arrive at the present crime scene and who took the early lead as chief investigator.

Smith, who had earned his bachelor's, master's, and D.V.M. from Ohio State, was teaching several classes in parasitology and assisting in Dr. Noga's lab, where, for his dissertation, he had been working on an experiment that looked at the immune response of fish to a parasitic marine dinoflagellate. Dinoflagellates are a group of microscopic, mostly single-cell organisms that belong in a "twilight zone" between the plant and animal kingdoms. Claimed by botanists as microscopic plants because some members obtain their sustenance through photosynthesis, they are also claimed by zoologists, because other members consume protozoans. A primitive group that has been traced by the fossil record back at least 500 million years, dinoflagellates form the base of the food chain in both fresh and salt water, being consumed by small organisms that in turn are eaten by larger organisms, all the way up to fish. For this reason, they are considered to be beneficial. There are, however, rogue species of dinoflagellates that can be offensive, producing toxins or, in the case of the dinoflagellate Smith was studying, causing infections.

In order to test the immune response of "naive" fish—that is, fish with no prior exposure to this particular dino and thus no chance of developing antibodies to it—Smith was working with tilapia, a small African genus that could survive in both fresh and salt water. He was conducting his experiments in a ten-by-ten-foot room lined with shelves on which five- and ten-gallon tanks were set, each fed by tubes that led from pipes connected to a central supply of aerated well water. He had begun, on a Monday, by putting fifty tilapia in a 300-gallon fiberglass tank filled with fresh water, after which he had gradually increased the salinity of the water by adding Instant Ocean salts. On Thursday he had raised the level to three parts per thousand, and on Friday he raised it to ten, full-strength salt water being thirty-five parts per thousand. On Saturday he came in just to make sure the fish were okay, and he found half of them dead and the other half dying.

His first thought was that something was wrong with the water-quality controls. Either the biological filter had gone down or someone had accidentally shut off the air supply, the two most common causes of fish mortality in an aquarium system. As for the cloudiness in the water, he assumed that was the by-product of fish beginning to decay.

He didn't really think much more about it than that, but after body-bagging the dead fish and before doing a water exchange to get rid of the cloudiness, he performed a water-quality analysis. Using a kit designed for just these situations, he checked for the usual environmental toxicants and was surprised to find that the ammonia levels were fine, as were the nitrite and the pH. Since he was unable to figure out what was wrong with the water by the obvious methods, and as there was obviously something clouding the water, he took a sample up to the laboratory and looked at it under a light microscope, where he found the specimen swarming with weird little organisms. That didn't strike him as particularly unusual—he knew that when fish died they served as a nutrient source for a lot of microsopic things that throve on destroyed tissue—but these critters had an unusual swimming pattern, similar in some respects to that of the dinoflagellate he was studying. So he took a water sample and spun it down in a centrifuge, and what he saw then did give him a start. In just a few drops of water, thousands of these organisms were present, and they were not the dinoflagellate he was familiar with.

Thinking that perhaps this was another type of parasitic dinoflagellate, he took some living-tissue scrapes off the dead fish and did the normal diagnostics. But after examining gill tissue and mucus scrapings, he did not find anything attached to these fish, which was the way a parasitic dino did its damage. Baffled, he decided to wait until Monday to report the massacre and pose a stunning puzzle to his adviser.

Before people met Dr. Noga, they thought perhaps he was Japanese— but it's an abbreviated Polish name. Although his interest in marine biology went back to a south Florida childhood, and an aquarium he'd stocked with angelfish caught while snorkeling off the beach and with sailfin mollies scooped from ponds at a local golf course, his appearance suggested a buttoned-down businessman. He was corpulent, with a round, fleshy face that made him look younger than forty. He combed and parted his brown hair neatly, and his idea of casual dress was a shirt and tie. After working under him for almost two years, Smith had ambivalent feelings toward his adviser. Smith was a team player, while Dr. Noga seemed to be oriented toward personal advancement—and was not above taking credit for work performed by those under him. While this irked Smith, he knew it was not an uncommon phenomenon in research universities. He had made a pledge to himself that he was going to be one of the first students ever to leave a graduate program and not complain about or despise his adviser. When he came to Dr. Noga that Monday, he was not thinking along these lines, however, because he had no sense that this was a big deal. His report was given in a spirit of: Gee whiz, guess what happened. What do you think might be going on?

Dr. Noga had no idea and wasn't all that bothered. When he suggested to Smith that he drain the tank and start the experiment over and maybe that would be the end of it, Smith agreed with the approach. But several weeks later, when he began to raise the salinity level again, he had another die-off. And when he put a sample of the cloudy water under the microscope, he saw the same little bugs swimming around.

Whatever was happening and whatever was behind the unusual mortality was something totally secondary to the research Smith was doing. But it was interfering with his primary work, so with Dr. Noga's permission, Smith proceeded to study it on the side, thinking he might get a small paper out of it, for submission to a minor publication.

He began with a few preliminary experiments that were not designed with controls but done more to see what happened. Filling a separate tank with the water containing the mysterious organism, he dropped in different species of fish that were around the lab. He found that striped bass were extremely sensitive. In a relatively short time they would begin to exhibit a stress syndrome—they would go off into a corner, they would start losing coordination and roll on their side and become inactive, they would begin to lose their ability to control coloration, and they would die. Goldfish, on the other hand, were hardier, lasting about as long as the tilapia.

He also noticed a relationship between the cloudiness in the water and its deadliness to fish. When he put in fish, a day or two later the water turned cloudy, and a day or two after that he was counting corpses.

At that point he made two observations: If he took the dead fish out and didn't put any more fish in, the cloudiness disappeared. And when he then looked at the water under the microscope, the dino seemed to have disappeared too. But if he put in more fish immediately after the first group died, there would be a rapid onset of neurological problems in the new fish, sometimes within a matter of minutes, suggesting that the cloudy water was hot with toxicity. And when he looked at that sample under the microscope, it was swarming with dinoflagellates.

Smith decided it was time to test for the presence of a toxin. To do this, he took three fifty-milliliter beakers, filled one with fresh sea water, another with water from the tank that potentially included the organism, and the third with that same water after it had been run through 0.2-micrometer filters to remove all possible sources of bacteria, as well as all dinoflagellates. Then he added five tilapia fry to each beaker and left them in an incubator for a couple of days. When next he checked, the fish in the fresh sea water that had not contained dinoflagellates were doing fine, all the fish in the water with the organism were dead, and where dinoflagellate-containing water had been filtered there was a 60 percent mortality. This proved, conclusively to him, that something the dinoflagellates were excreting into the water was lethal to fish. And in case there were doubters, he redid the experiment twice, so he could get actual data and numbers to back that assertion.

Still thinking that the end result of all this would be a minor scientific paper, Stephen Smith went to the literature to find out if anything like this organism had been described. He turned up very little, so he started calling the people listed in the reference sections. While he had several interesting conversations, when he offered to send samples he had very few takers. And because he did not personally feel he was in a position to carry the research much further on his own, both he and Dr. Noga began calling people within North Carolina whose disciplines touched even tangentially on dinoflagellates.

When she first heard from Dr. Noga, JoAnn Burkholder thought, *This is all very interesting, but why is he calling me?* She was a plant person, not a fish pathologist. Furthermore, although Dr. Noga seemed pleasant enough on the phone, she'd never met the man, and it crossed her mind that maybe the word had spread that she was gullible when it came to helping other researchers with their work.

"That's very interesting, Dr. Noga," Burkholder commented, "but I'm not sure what this has to do with me."

Noga came to the point. "I was wondering if you'd be willing to take a look at this and tell me what you think. We believe we have a toxic dinoflagellate, but we're having trouble identifying it."

Dinoflagellates had been part of Dr. Burkholder's general algal studies, and she had attended many seminars, some of them featuring the leading researchers on toxic dinoflagellates. What she'd learned had fascinated her. Of the several thousand dinoflagellates that had been identified, some two dozen species were known to produce toxins; and although millions of dollars had gone into research on them, they were still poorly understood. Despite their tiny size, the damage they were capable of causing gave them an almost frightening aura. When conditions were right, dinoflagellate populations could explode astronomically, forming dense, colored blooms; and if the species undergoing this sudden multiplication happened to produce toxins, the results could be catastrophic. "Red tides," as these blooms were called, poisoned the water. Fish were killed by the millions, as sometimes were birds that fed on them. While shellfish often managed to survive, they frequently became contaminated, and it was through them that humans became seriously ill.

Their deadly legacy was the stuff of legend and lore. The first record of a red-tide bloom appeared in the Bible: " . . . and all the waters that were in the river were turned to blood. And the fish that were in the river died; and the river stank, and the Egyptians could not drink of the water of the river; and there was blood throughout all the land of Egypt" (Exodus 7:20–21). When Captain George Vancouver and his crew landed in British Columbia in 1793, the Inuits told him that they did not eat shellfish after the tides glowed at night, and it was later ascertained that the glow was caused by blooms of a phosphorescent dinoflagellate that produced an alkaloid so potent that a pin-size amount could kill a human being.

Red tides baffle marine scientists for several reasons. While researchers knew what they were, they didn't know what caused them, how to predict them, or how to stop them once they'd begun. What had become apparent to many scientists, however, was that they were increasing in frequency around the world.

Though Dr. Burkholder brought to the discussion a background that included a certain familiarity with dinoflagellates—enough for her to know, for example, that it was highly unusual for toxic dino-flagellates to cause problems in aquarium fish populations—she felt she was in no position to get involved in this matter. Glancing at a pile of papers on her desk that were calling for her attention, she said, "I'm sorry, Dr. Noga. I wish I could help, but I'm over my head with work as it is. I simply don't have the time to take this on. I wish you luck."

Two weeks later, Noga called again.

"I thought maybe you would reconsider," he said.

She was surprised. "What made you think that?"

Then it came out: He had approached four or five other scientists, but for one reason or another, none of them were interested in helping him.

Oh, so you thought I might be the exception? she thought. After listening, she turned Dr. Noga down once again. "Look, my back-ground is fresh water. I'm an ecologist, not a taxonomist. I've never really worked with dinoflagellates, so believe me, I'd be of little help."

A week went by before the next call came, and this time Dr. Noga said that they had got some pictures of the dinoflagellate under the microscope; would she please look at them and tell him what she thought? "We don't really know what this organism is, and we'd like to know if anybody has ever seen it before."

While nothing had caused her to change her mind between this phone call and the first, it occurred to Dr. Burkholder that perhaps she should at least acknowledge Noga's outreach efforts. Typically, science was compartmentalized. People who worked with fish diseases often didn't know a lot about algae, and algae people knew little about fish diseases. Each was involved in his or her specific area of research, and with science becoming more and more specialized, this was an increasing tendency. So the typical fish person would have been inclined to say, All right, we don't know what this is, but it's probably a virus of some kind; let's quarantine the new fish and start over. End of story. That Dr. Noga was asking for her help suggested a willingness to draw on other disciplines, which in principle she thought was a good thing. So this time, with an exasperated sigh, she relented. "Okay," she said. "Send them over."

When she saw the photographs, she was even less enthusiastic about getting involved. She knew that taking good, clear pictures of tiny organisms through a microscope was a task that required special preparations and procedures. And it was painfully obvious from what she received that neither Dr. Noga nor his assistant had mastered the art. It was true that the images resembled small algae, but the organisms were squashed and dehydrated, making identification nearly impossible.

There wasn't much she felt she could do with the photos, and that was what she told Dr. Noga when next he phoned.

"Well, do you know anybody else in the field you could ask?"

He was sounding desperate, so she finally said, "There are some people in Florida I've heard about who have a special interest in dinoflagellates. I don't know them personally, but if you'd like, I'll give them a call."

"I'd like," he said. "Very much."

The next day, she put in a call to the Marine Research Institute of Florida's Department of Natural Resources. She never got through to Dr. Karen Steidinger, the director, a world-renowned authority on dinoflagellates, but she did talk to one of her assistants. Even though she got the distinct impression that they were not excited about this, she was told to send them the photographs and several batches of live culture, and they would look at the material.

Months passed, and nothing came back from Florida. When Dr. Burkholder called for an update, she was told the culture had died; could she send more? She did, and the next time she called, when she asked, "Can you tell us anything about it?" she was told they were having trouble identifying it but thought maybe it was not just one dinoflagellate but three or maybe four different species. She got the impression that they had taken only a cursory look at the organism and were guessing.

Meanwhile, Dr. Noga kept calling back and trying to pique her interest. "Listen, I know fish pathology isn't your field, but algae isn't mine. There are a lot of neat things that could be done with its ecology, like figuring out its life cycle, how it responds to environmental conditions . . . what it is."

At the time, JoAnn Burkholder's laboratory was pitifully equipped. It was located in a room next to her office, so small that there wasn't enough space to allow for the installation of a fume hood. She'd been given $20,000 in setup money, which is nothing when it comes to science. It barely covered the cost of a modest microscope. She was unable to afford a technician, she had no student help, she had to do everything herself, and the bulk of her time was spent writing grant proposals to raise money.

She was also teaching a course in phycology—the study of algae. It so happened that in the spring of 1987, for the first time in history, an expatriate red tide had moved north from Florida on the Gulf Stream. It broke off and was carried to the North Carolina coast, killing fish, poisoning shellfish, and causing millions of dollars' damage to the state's seafood industry and tourist trade. Wherever possible, she tried to incorporate events in the real world into her lectures, so she had talked at length about the microscopic villain that was responsible for red tides; she even described to the class the weird dinoflagellate that seemed to be killing fish at the vet school.

As part of the phycology class, Dr. Burkholder required students to take on a special project. They could do anything they liked—a field survey, an experiment, a literature search on some aspect of algae that she had covered in insufficient depth for them. After class one day, a student of hers named Cecil Hobbs, a quiet, conscientious fellow who, at forty-one, had come back to school for his master's, approached her and said he had grown up on the North Carolina coast, where, for generations, his family had made their living raising oysters. But they had been totally wiped out by the red tide, and this had given him a personal interest in dinoflagellates. "For my special project, I was thinking maybe I could work on the ecology of that dino at the vet school."

Even as she heard herself reply, "I don't know. Let me talk to Dr. Noga," she was thinking, *What am I doing, getting involved in this?*

3

ecil Hobbs had been a high-school biology teacher for seventeen years before burning out on kids and going back on active duty with the National Guard. A long-range plan to get a master's degree also figured into his new future, which is how he came to be a student in Dr. JoAnn Burkholder's phycology class. Algae had not been a special area of his interest—his specialty was entomology, the study of insects—but that changed during his semester in Dr. Burkholder's class. Quite simply, he thought she was the best college professor he'd ever had. She loved her work, believed it was important, and had the ability to present it to her students in a way that could excite them.

Relatives of his who had made their living on clams and oysters had indeed been wiped out by the red tide. Poor people to begin with, they had lived from catch to catch, and now they were dependent on food stamps. That certainly gave him an incentive to play detective as his special project, but it was Dr. Burkholder's lectures about these intriguing creatures that really hooked his interest. Looking at them in class under a microscope had been like an aerial reconnaissance of a primitive world not yet fully explored, and he had found dinoflagellates as fascinating as some people found dinosaurs.

The arrangement worked out by Drs. Burkholder and Noga authorized Hobbs to go to the vet school and run very simple ecological experiments. This was not a full-time project but rather something he undertook on a casual basis. The vet school was near the National

Guard post, so he would stop by during the lunch hour or after five to dabble with the culture. The very first thing Burkholder had instructed him to do was repeat the experiments Noga and Smith conducted, because she had never heard of an organism behaving as they had reported, and she wanted it corroborated. Hobbs found that Smith and Noga's experiments were demonstrable and repeatable. When fish were added to the dino-infested water, they would die.

Next Dr. Burkholder told him that before subjecting the organism to any tests, he should watch it under the microscope, get familiar with what it looked like, and, when he was ready, prod it with simple tests that betrayed its habits. Hobbs discovered that when he took a drop of water and put it under a light microscope, there were so many starlike organisms swimming around, he had trouble recognizing the dino-flagellate he was supposed to be studying. It was such a basic step that he was reluctant to admit his difficulty to Dr. Burkholder. When he finally did, she laughed and told him to bring a sample back to her lab; they would look at it together.

She wasn't laughing, however, when she put the sample under magnification and brought it into focus. Indeed, she sat for a long time watching the activity in the water, because she too was confounded by what she saw. One of the characteristics typically used to identify dinoflagellates was size, but the size range among these was incredible. The largest were almost ten times bigger than the smallest.

"Are they all dinoflagellates?" he asked.

"I don't know what they all are," she had to admit.

Another characteristic of dinoflagellates was that each species had a distinctive shape; yet all the organisms she was looking at were similarly configured. "They look like the same thing, only miniature or bigger."

Several more minutes passed before she spoke again. "Cecil, I want you to count everything and break down your findings into size categories. Count the little ones and the big ones, and make notations about what you're counting, so we can backtrack once we know what's going on."

She had one other suggestion: She steered her graduate student toward the literature that would allow him eventually to take quality photographs with a camera mounted on a high-powered scanning electron microscope.

Over the next six or seven months, Cecil Hobbs conducted a series of basic experiments that were designed primarily to determine how the organism in the culture responded to changing environmental conditions. He already knew it was motivated by the presence of fish, but was it more active in fresh or salt water? Did fluctuations in temperature inhibit or stimulate it?

The majority of his time, though, was spent in perfecting the techniques that would allow him to capture what was going on in photographs taken with a scanning electron microscope, a very difficult endeavor. There were specific procedures for killing microorganisms so as not to disturb their primary characteristics—the faster the kill the better, for example, because slow kills destroyed cells. Then there were special ways of preserving the specimens in a natural state, and methods of preparing them on a slide—all of which took hours of trial and error. Getting good shots was an art, not a science.

One afternoon, Cecil Hobbs walked into Burkholder's office and dropped into the visitor's chair. She knew he was frustrated by his slow progress with the organism, but on this occasion he looked thoroughly depressed.

"What's wrong?" she asked.

He took a deep breath. "There's been an epidemic at the vet school."

"What does that mean?"

"Everything is dying. The fish in my tanks, the fish they were growing in the fish barn, all the way to the fish in the display aquarium in the front office."

"When you say dying . . ."

"The symptoms are the same as with the tilapia."

"Did you take samples?"

He nodded. "Dr. Noga gave me samples from each of the tanks."

"And?"

"I found the same dino in all of them."

Her mind racing for some sort of explanation, Burkholder recalled a conversation she'd had with Dr. Noga early on, in which she warned him to isolate the aquariums in which the toxic dinoflagellate was known to exist, and to make sure nothing was done that could spread the contaminant from one aquarium to another. "Did they do what we told them? About quarantining?"

Hobbs acknowledged her suspicion with a short shake of his head. "Not really, and I'm afraid that's what might have happened. He has students who come in on weekends to monitor the fish, and I think they were using the same refractometer when they went from tank to tank, checking the salinity levels."

You didn't have to be an expert to know you should take precautions when dealing with a toxic dinoflagellate. Dipping the same refractometer in tank after tank was tantamount to taking the temperature of someone sick with flu, then sticking the thermometer in the mouths of a dozen healthy people.

When Cecil Hobbs returned to the vet school, Dr. Noga asked him into his office. The risk of this organism's continuing to infect healthy stocks was so great, said Noga, that he had ordered all his tanks disinfected with Clorox, and he was calling off all further research. At this point he didn't care what the organism was; he just wanted it gone.

When Hobbs broke the news to Dr. Burkholder, it included a personal appeal. He felt as if he was on the trail of this killer and close to understanding its modus operandi. Could he take the culture home with him and continue studying it there on a temporary basis?

"That's okay with me," she answered. "And you can bring samples into my lab if you want to work on them here."

"I'd like to do that," Hobbs said.

After clearing the idea with Dr. Noga, Hobbs drove to the vet school in his Ranger and loaded two ten-gallon tanks into the back. One had the organism in it, the other was stocked with tilapia fry. He then drove the four miles to his house and one by one carried the tanks down to his utility room in the basement, where he put them on a bookshelf beside the washing machine and the freezer.

Activities at the National Guard preoccupied Cecil Hobbs for the next several months, so he did little more than feed the organism every two or three days—and when the fish died, he flushed them down the commode. This was not something he enjoyed doing—he liked fish and found himself becoming attached to the tilapia fry—but he continued it in the belief that this was what it took to keep the organism alive while he waited for an opportunity to resume serious research.

The sequence of events that followed causes him even today to shake his head in disbelief.

Often, in the evenings, he would talk about his project with his wife and family, so they were aware that he had brought something into the house that demanded the ritual sacrifice of fish, and it was living in their basement. But it never occurred to him that his oldest daughter, Lisa, would respond the way she did.

Lisa, the product of an extremely difficult birth, was mentally handicapped to the degree that she was homebound. During the day she liked to sit and watch the fish swim round and round in the big aquarium her father kept in the living room. It was hard to know how much Lisa, with an IQ in the fifties, understood of what went on around her. Cecil Hobbs knew this, of course; but that wasn't the way he was thinking the day he came home and went downstairs to the utility room to feed the organism and found the tank sitting empty on the shelf.

For a moment he stared uncomprehendingly. At a sound from behind, he turned and saw Lisa watching him with a concerned expression.

"Lisa," he said, his breath coming fast, "what did you do?"

"Daddy, that water is bad," she said.

He grabbed her shoulder. "*What did you do?*"

"I poured it out, Daddy. It was killing the fish."

Six years had passed when Cecil Hobbs related this incident to me, but I could feel the emotion that made him tremble with rage that day. To his knowledge, there was no culture left at the vet school; the only water known to contain this organism had been in his possession, and now that had disappeared, literally down the drain. All the work he'd done over the previous year was wasted.

"I won't tell you what I said to Lisa," he murmured to me, as if he preferred to forget the whole thing.

I nodded sympathetically. "I can imagine how you must have felt." But I also wanted to know what happened next. "Would I be out of line if I asked your daughter what she remembers?"

Lisa is in her twenties now, but on the phone she sounds about seven or eight, and very sweet. "It was a accident," she told me when I asked her if she recalled that day.

"I'm sure it was, honey," I assured her.

"My daddy got mad."

"I know. He told me. Can you tell me what you did that made him so mad?"

Bit by bit, the story came out. For weeks she had seen him feed the fish and watched them die. It made her sad. Finally it had been too much for her to bear. She wasn't trying to save the fish. They were already dead. She just didn't want what was killing them in the house anymore. One by one, she took the dead fish out and flushed them down the commode, just as she'd seen her daddy do. Then she carried the tank full of bad water into the backyard and dumped it out.

"You did a good thing, trying to help the fish, Lisa. Thank you for talking to me."

The conversation ended with Lisa reminding me, "It was a accident."

In the days that followed, Cecil Hobbs was so discouraged that he considered calling Drs. Burkholder and Noga, telling them what had happened, and dropping the project entirely. But he didn't, and it was because something in the back of his mind wouldn't let him give up just yet. It was as though he was waiting for a sign—and then it came; the thought that if this organism could spread so easily from one aquarium to another, perhaps it was hardy enough to survive on the walls of an empty tank. He decided to test it. Not wanting to use tap water because of its chlorine content, he drove back to the vet school, where he filled a bucket with water pumped from a well on the premises; returning home, he poured the contents into the tank. Adding several tilapia, he crossed his fingers and waited.

The answer came in two weeks. He was as amazed as he was relieved. Floating fish let him know the killer was back.

Shortly afterward a space opened up on the North Carolina State campus in a large structure adjacent to Dr. Burkholder's office and lab building. Called a phytotron, it was dedicated exclusively to growing and experimenting with plants. Hobbs set himself up there, but because the conditions were so good for algae to grow, species unrelated to his work began swamping his tanks. Feeling as if he was running out of options, Hobbs approached Dr. Noga to see if there was any chance of returning to the vet school. As it turned out, the timing was perfect. There had just been another epidemic of fish killed at the vet school, alerting Noga that even though no one was working with an active culture, the organism was still getting to his fish somehow. Realizing it was in his best interest to find out once and for all what the hell this thing was, so he could find a way of controlling it, Dr. Noga gave Hobbs permission to work in an isolated environmental chamber at the school, and Cecil Hobbs was back in the research business.

He'd been dabbling with this dinoflagellate for almost two years now, with little to show, but over the next six months his efforts began to pay off richly. Previously, he would on occasion bring Dr. Burkholder a photograph he'd taken under the scanning electron microscope; she would interpret what she saw, but not much could be deciphered because the pictures had not revealed a lot of detail. Now that he was beginning to master the photo process, the pictures were much clearer—even though he didn't always understand what he was shooting. He amassed a series of sharply detailed photos taken of the dinoflagellate at a range of times and under a variety of conditions, then delivered the batch to Dr. Burkholder for analysis.

Burkholder looked through the photos rapidly, then more slowly. And to say it was as though she were seeing this organism for the first time is only half the story. She had the feeling that what she was looking at, *no one* had seen before.

Asking questions that would connect the photos to the time line of when fish were added and when they began to die, she was able to piece together the life cycle of this elusive creature. And the story the photos narrated could scarcely have been bettered by science fiction.

Left alone, this microscopic alga seemed content to rest in a dormant state in the sediment at the bottom of an aquarium, encrusted in a scaly shell, where it appeared to thrive on photosynthesis, as you would expect from vegetation. But as soon as fish were introduced into the water, it somehow sensed their presence, and a bizarre series of transformations was initiated that turned a harmless little cyst into a veritable sea monster.

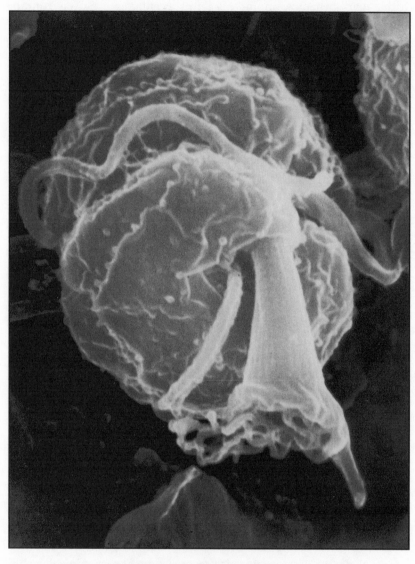

The killer-dinoflagellate, *Pfiesteria piscicida,* in its most toxic stage.
Photo by Howard Glasgow.

Bursting out of its shell, a swimming cell propelled itself through the water by means of two whiplike tails, or flagella. Then, as it rose through the water column and approached a fish, it apparently unleashed a powerful neurotoxin that first stunned and eventually suffocated its prey, evidently through muscular paralysis. This toxin seemed also to have a corrosive effect on fish skin, which, as the victim died, would slough off. At this point the dinoflagellate adopted yet another form, enlarging in size to accommodate the growth of a tonguelike absorption tube, called a peduncle, which the dino attached to fish flesh and fed through at a leisurely pace.

There was an epilogue. If at any point during this blitzkrieg the fish were removed, the organism sounded retreat, dropping back to the bottom of the tank, where it again donned the bristling armor of the cyst stage and waited in suspended animation until more fish came along. In which case it went on the attack again.

No form of life JoAnn Burkholder had ever heard of was supposed to behave like that. Certainly not a single-cell aquatic plant. The toxic dinoflagellates she was aware of emitted toxins to outcompete neighbors or deter predators, but none—not one—behaved with such predaceous purposefulness.

It was macabre. It was fascinating. This explained the size differences they had seen under the microscope. It also accounted for the difficulties in fingering the villain behind the epidemics at the vet school: by the time investigators arrived at the crime scene, the face of the killer had changed.

What it did not explain, however—and the question became preeminent at this point—was: Where did this thing come from? Was it something brought in accidentally on one of the exotic species of fish in the laboratory? Burkholder knew Dr. Noga had brought in fish from places as distant as Australia and Israel, which meant an import from anywhere in the world could have colonized in the vet school's tanks. But Noga also worked with a lot of local fish—spot, croaker, flounder —drawn from North Carolina's estuaries.

What they had on their hands was a novel culture contaminant of unknown origin. As far as she knew, this organism wasn't on the map in terms of any natural setting. And even though JoAnn Burkholder was intrigued, she was in no position to mount an aggressive search for it in the wild. This whole endeavor had been just a sideshow to the main events that occupied her time: sea grass and reservoir studies, teaching botany courses, the effort to get grants funded.

So she phoned Dr. Noga, with whom all this had started, to see if he had any thoughts on its home address.

"Have you ever tried to track it?" she asked.

"Well, we've thought about that."

If he could have seen the scowl on her face, he would have done a lot less hemming and hawing. "But you haven't."

"No, we haven't."

She thought that was curious. She would have expected that the person impacted most heavily by this organism would have been anxious to trace its origin. It had been hanging around his lab for going on two years.

Something clicked in her mind, as she recalled that in the North Carolina press recently there had been reports of fish dying in large numbers in the estuaries. Maybe there was no connection, but there was no harm in checking it out. It was certainly a lot closer than Australia.

"Have you tried talking to state biologists to see if they've seen anything that resembles this in the estuaries?"

"Ah, those guys are hard to work with," Noga replied. "I've tried to get them to help me by going out and sampling fish kills, but if it's a bad-weather day they won't even float their boats."

Burkholder's patience with Dr. Noga was wearing thin. She didn't know him well enough to understand his seeming indifference; it could have been that as a fish pathologist, he just wasn't inclined to extend himself beyond the vet school. But she had been generous in donating time and effort to what essentially was his cause, and it annoyed her no end that he did not seem to be interested in following through on the obvious last step.

Nevertheless she had brought it along this far, and now she asked for names. Even though they wouldn't know who she was, she intended to put the word out to fishery biologists: If you see a fish kill, please collect water samples and send them to us as fast as you can. Or call us and we'll come down and get them.

Although she didn't have proof and hoped she wouldn't be called upon to make a case for the claim, to get their attention she added: We'd like to try and help you find out what's killing the fish, and we think we have an idea what might be doing it.

4

It was only mid-May, but to Kevin Miller, loading his essentials into a nineteen-foot Glassmaster, it felt like the start of an August scorcher. One of those days that took off with a hot dawn toward a record high before the sun went down. By the time he motored away from the dock, Miller was ready for a quick dip in the river, but instead he settled for the breeze that blew up with a turn of the throttle. Behind him, the historic waterfront of two-story eighteenth-century brick buildings called "Little" Washington—the name locals used to distinguish their quaint colonial North Carolina coastal community from the nation's capital—glowed with a solar aura in the rich morning light.

Although his massive stature, wide girth, and full black beard gave him the look of a mountain man, Kevin Miller was a water boy. In the baby book kept by his parents, the first question they recorded him asking was: Where does the water live? Two years old then, he was in his mid-thirties now and working as an environmental technician in the water-quality division of a regional branch of North Carolina's Department of Environment, Health and Natural Resources. More pointedly, he was the department's representative to PERT, the acronym for Pamlico Environmental Response Team, a crack group of researchers and investigators formed by the governor of North Carolina in reaction to a public outcry over what locals and the press were calling "the poisoning of the Pamlico River estuary."

The Pamlico was wide and sluggish at this point, the shore edged with brackish marshlands. To almost anyone else, the view of wild rice and seashore mallow, red cedar and bald cypress would have confirmed North Carolina as a picturesque state that showcased nature in all its splendor—but as Kevin Miller's eyes played over the water, they were scanning for ominous signs.

For over a decade now, the Pamlico River and its main tributary, the Tar, had been a body of water plagued with unusual events: an uncommon number of large fish kills. Sudden die-offs by the millions—usually of menhaden, a small, oily fish of commercial value, and a dietary staple for larger fish. A fish kill, in itself, was not all that unusual in North Carolina's estuaries, and the common theory that everyone had been comfortable with for a long time pointed to low dissolved oxygen as the cause. The dynamics of "low DO" went this way: In the warm summer months, algae blooms would occur in the estuaries, where the fast-running inland rivers were slowed into lakes by the tidal inlets. Algae, like any green plant, produced oxygen during the day but consumed it at night. Bring a school of fish into an area where the oxygen was depleted, and bingo, you had a kill.

There was only one problem: The theory didn't account for the dramatic increase lately in the number and size of the fish kills. Nor was low DO present at some of the larger kills. As Barry Adams, another member of PERT, tells it: "This one time I was called down on the river because fish were flipping right there in front of everybody, and by the time I got there, three TV cameras were set up, waiting to film me. So I took out my instruments for dissolved oxygen, salinity, and pH, and I ran all the tests, and it was intensely embarrassing, because all of my meters were saying nothing's wrong."

Nor did low DO address some of the strange phenomena taking place on the Pamlico River—the strangest of all being reported by crabbers out checking their pots who would come upon a pier or piling or net stake and see crabs by the hundreds, trying frantically to climb up in an effort to get out of the water.

Everyone from the area was familiar with "fish walks" and "flounder jubilees." When dissolved oxygen levels were especially low in the river, fish—particularly flounder—would sometimes move into the shallows along the shore where waves were breaking and re-aerating the water. It would look as if they were trying to walk out of the water, but they were just trying to get their oxygen, and they would be so exhausted that people would walk up and down the beaches, filling their coolers.

Crabs were also known to do a "walk," so at first folks thought maybe this was why they were trying to climb the pilings. But two things were different about this situation. When technicians from the state came out with their instruments and took measurements, the dissolved oxygen levels were normal. Second, these crabs were anything but lethargic. They moved as if they had a fire under them.

This was the combination of events that had led to the formation of PERT. But for all the expertise and alertness that had been brought to bear on the problem, its members, after two years of collecting information and documenting fish kills, were still scratching their heads. The best that could be said of their effort was that it had improved the way the Department of Environment, Health and Natural Resources did business. Before this, the protocol said you waited until a complaint was lodged before acting. Somebody would wake up in the morning and notice fish washed up on the beach or some nasty brown scum on the water, and they would call the 800 number for the PERT team, which swung into action. Rushing to the scene, they would scoop bottles of water for lab analysis, survey the species composition of the fish, and do a rough count.

But by the fall of 1990, the team realized that this way of going about investigating these fish kills was wrong. They had come to recognize the kills as dynamic; if they truly wanted to understand what caused them, they needed to be more systematic in their approach, because typically, menhaden fish kills didn't happen in one area and last a day or two and end. They went on for weeks, and they weren't concentrated in a single spot. You saw dead fish in a certain part of the river one day, and in another part the next day.

This had led to a revised strategy. From then on, whenever a fish kill was in progress, rather than chasing down dead fish the technicians would visit established stations along the river—a navigational marker, an eyeballed spot halfway between two creeks—and take water samples and make environmental observations. This explained Kevin Miller's outing on that sweltering morning in May: a menhaden fish kill had been going on for about a week.

It was a distance of about thirty miles between Little Washington and where the river widened into Pamlico Sound, and whereas Miller normally went down one side of the river in the morning and came back up the other in the afternoon, taking a full set of samples at every station, he decided to vary his routine. He was looking for the worst station on the river that day—the place with the largest number of fish in distress or dying—and there he intended to take the most complete set of measurements his kit allowed. He also had decided not to hit every station but just to check three or four sites in the morning and then return in the afternoon on the chance that conditions had changed.

By late afternoon Miller had gone all the way down the river and was halfway back up, and he was still looking for that one worst station. He was also running low on fuel, so he turned into a creek that led him to a small marina at the town of Bath. After gassing up, he came back out into the main channel and for the first time noticed a black mass of clouds gathering into a thunderstorm to the south.

Since afternoon storms usually moved straight across the river, Miller didn't give it much thought as he steered toward the next station, which was across and upstream just a bit. But this storm did something different: it turned and followed him.

Although he noted the meteorological quirk and could see and smell rain falling less than a mile away, it was warm and sunny where he was. Arriving at the station, a shallow indentation just downstream from a major point that jutted into the river, he went about his work at a normal pace. He had just finished taking a wind-speed measurement—it was about five miles an hour—and was preparing to chart the dissolved oxygen level, when he realized it was much darker and colder, and the wind had picked up.

With the storm almost on top of him, he yanked up the probes and sampling equipment, about to make a run around the point into the lee. Then, just out of curiosity, he held up the wind-speed meter, in time to record a gust of forty miles an hour.

As it turned out, he no sooner rounded the point than the storm turned inland, missing him completely. Looking at his watch, he wondered if he should return to the station he'd abandoned. Based on what he'd seen, it was no more promising than it had been that morning. But he knew that a river can be physically affected when a storm passes over it, so just to be on the safe side, he headed back to finish what he'd started.

A changed scene awaited him. The storm had roiled the water, and where there had been nothing unusual before, now there were fish in distress all over the place: menhaden swimming with their tails down, which was extremely unusual, and rising to the surface and gulping air as if they couldn't breathe through their gills. He also noticed that the water was cloudy. Without hesitating, he started taking collections and measurements.

It was quite late, after seven, by the time Miller docked in Little Washington; and later still by the time he had returned to his office, completed the paperwork, and drawn a map to accompany the field samples he'd collected. He stuck them all in a cooler packed with vials and bottles and ice, which he set out for the state courier system to take to Raleigh. There, it was to be delivered to a North Carolina State scientist by the name of Dr. JoAnn Burkholder. At the time, Miller knew he had *seen* something unusual, but in his mind it was the eerie way the storm had seemed to track him. As for whether he had *found* something unusual, all he knew was that he'd discovered the station he'd been looking for.

~~ ~~ ~~

Several days later—on May 21, 1990—JoAnn Burkholder was sitting at her office computer, writing a paper. She had personalized her professional quarters much as she had her apartment. Her idea of a No Smoking sign was a cartoon from *The Far Side* taped to her office door: several brontosauruses puff heavily on cigarettes, above the caption "Why Dinosaurs Are Extinct." The clutter behind the door—bookshelves and filing cabinets doubling as tabletops for tottering stacks of paper—suggested that the occupant had a dozen things going at once. Rock-concert ticket stubs tacked on a corkboard—Eric Clapton, Pat Benatar, Rod Stewart, the Rolling Stones at RFK—and a large photograph of Jimmy Page, lead guitarist for Led Zeppelin, revealed more than her taste for music: they constituted a confession about what she did to relax.

As she wrote, she so immersed herself that the ringing telephone startled her. She grabbed at the receiver, as much to quiet the phone as to answer it. "JoAnn Burkholder."

It was Cecil Hobbs, and he was calling from her lab, which had recently been relocated to the basement. "Hey, JoAnn. Can you come down here for a minute. There's something I'd like you to look at."

Her eyes had never left the computer screen. "Is it really important? I'm in the middle of something."

"I think it is. I think we've finally got this dinoflagellate in a field sample."

The spell snapped. "No! Really? Where did it come from?"

"A fish kill on the Pamlico River."

She raced down the hall and took three flights of stairs two at a time; it was all she could do not to fling open the door to her lab.

Cecil Hobbs was standing beside a microscope, a grin on his face. "Take a look at this."

She looked, and there was no question. It was the dino they had been working with in culture since the murder at the vet school, almost three years earlier.

Caught in the act of killing fish, it was still at its most lethal stage. "Well, well, well," she murmured, watching it swarm. "I've been waiting to meet you for a long time."

Just to be sure, she poured the contents of the vial into an aquarium and added three fish. . . . Soon she was watching them do their watery death dance.

PART TWO

 "Oh, science took it awfully well."

5

For Dr. JoAnn Burkholder, changing an unknown and novel culture contaminant to the status of an estuarine pathogen was the equivalent of a scientific breakthrough. When she thought about why no one had known about this type of organism before, it didn't surprise her that she was the first. Its ephemeral nature made it easily missed in routine phytoplankton examinations. It changed into so many forms so quickly; it was colorless; and when it bloomed, it did not discolor the water the way a red tide did. You would almost have had to be looking for it in order to find it.

It was personally satisfying that she had been a good sleuth for this case, with her intuitive perception that the dinoflagellate might be what was killing fish in the estuary, and with the tracking skills that figured out you needed to be there while fish were dying in order to catch it. And though at this point she had no idea where an ongoing investigation would lead, she did know it was impossible for her not to want to pursue the answers to follow-up questions.

Where did this organism fit in terms of the classification of known dinoflagellates? She thought it was probably a very old species, because it resembled a genus of the oldest known dinoflagellates. She also believed it might be a missing link, representing an evolutionary departure toward a more complex dinoflagellate.

For a moment she let her mind run with that thought and tried to imagine the environmental pressures that had guided the dino's evolution. Multicellular organisms were generally considered more advanced, more sophisticated, more complex, than unicellular organisms, because their composition frequently represented a collection of highly differentiated cells that were organized to perform very specialized jobs. What was incredible about this organism was that a single cell had evolved such an astonishing array of mechanisms for surviving a wide range of environmental factors. It appeared to thrive in fresh water as well as salt; take away all water, and it still managed to stay alive. It fed on fish and, when there were none, hibernated until they came back. Most dinoflagellates were pretty much one-pathers—their behavior could be clearly charted, their responses more or less predicted—but this one had a bag of tricks. It was truly a marvel of nature, and Burkholder felt she had only begun to unravel its concealed powers.

She had other questions: What did fish make the mistake of secreting that caused this dinoflagellate to launch an attack? What was it doing to fish at sublethal levels? Did it have an effect on a fish's ability to reproduce, for example? But the burning question for her at this particular moment was: How important a factor was this organism in other fish kills? Sure, they'd caught it at one kill, but was it present in *all* the fish kills that were reported? Half? A third?

For an answer, she reissued her request to state biologists, asking them please to continue to collect and send samples from fish-kills-in-progress. She added that if they could notify her in time, she would come to the scene personally. She badly wanted to see for herself whether fish in a kill in the wild were exhibiting the signature symptoms she'd witnessed in the lab. She also wanted to get photos of fish dying, because she knew that documentation was going to play an important role in validating her research findings for others.

Seventeen major fish kills were reported that summer, affecting an estimated one billion fish. On two of them she was informed in time to jump into her 1980 Toyota Corolla and race to the coast. But in each instance she arrived too late. At one, the wind had blown all the dead fish out to sea; and by the time she got to the other, the kill was over and bulldozers were scooping fish carcasses off the beach and into dump trucks.

Nevertheless she did receive water samples from all seventeen kills, and under the microscope she found the organism present in half.

That could mean low DO was the cause of half the kills, or it could mean the organism had receded by the time the sample was taken. And still missing was proof that it was the causative agent in those kills where its presence was established. Still, to find it hanging around at the scene of 50 percent of the kills was a significant statistic.

A logical step at this point was to check the historical record on fish kills in the state, and for that Burkholder contacted the North Carolina Department of Environment, Health and Natural Resources in Raleigh. She knew that within the environmental agency there was a biomonitoring section, which for years had received field samples sent in by biologists from rivers and estuaries throughout the state.

By phone she reached a young woman by the name of Karen Lynch, who said yes, she was in charge of analyzing the phytoplankton samples and identifying and characterizing the algal content and count; and of course she'd be glad to stop by N.C. State and take a look at some pictures.

The next day, in a darkened second-floor teaching lab, Dr. Burkholder projected a slide of the organism as it had been photographed by Cecil Hobbs. "Karen, in the samples you have received over the years, have you ever seen anything that looked like this?"

She was expecting the young woman to squint, frown, ask if there were other photos, from better angles. But without the slightest hesitation, Karen Lynch said, "Yeah. That's *Gymnodinium aurantium.*"

Burkholder, who had been leaning against a table, stood suddenly straight. "What?"

Karen Lynch repeated herself.

Stunned, Burkholder said, "Are you sure? You've got to be *sure.* I'll show you a few more pictures."

She brought out a half-dozen photographs of the organism in its most toxic and lethal state, and after looking at them all, Lynch said, "There's no question. *Gymnodinium aurantium.* That's what we call it."

"Where did you get that name? I've never heard it before."

It was a bogus name, it turned out, invented by a bearded, violin-playing hippie biologist at the University of North Carolina in Chapel Hill, who had completed a doctoral dissertation on estuarine dinoflagellates almost twenty years earlier, before dropping out of academia. After noticing the organism in water samples at fish kills, he had been unable to find any description of it in the literature, and rather than going through the tedious formal naming process, he made a name up.

"And you've seen it in samples brought in by fishery biologists down on the coast?"

"Absolutely."

"For how long?"

"As long as I've been analyzing phytoplankton samples. Since 1985."

"And where have these samples come from?"

"The Pamlico River. The Neuse River, near the Cherry Point Air Station."

"Karen, I want all your fish kill records. As far back as they go. And I want them as soon as you can get them."

They arrived within a week. And when Burkholder checked through the files, there it was: The documented presence of the organism nick-named *Gymnodinium aurantium* in the phytoplankton count at fish kills with unknown cause dating back to 1985—in some cases, in numbers far higher than the lethal concentrations she'd worked with in culture.

Oh, man, she thought. *This fits perfectly.*

Adding a historic database to what she'd gathered from the fish kills tracked that summer strengthened the case, in Burkholder's mind, for implicating the organism as a causative agent in many of North Carolina's fish kills over the previous decade at least. But Burkholder also knew that the standard for scientific acceptance of such a claim was very strict and would require more study and substantially more evidence.

The problem now became one of funding. Where was she going to go for financial support to underwrite further research? She was an algal ecologist. She had no background or training in fish pathology or toxic dinoflagellates, no pedigree in the area she wanted to explore. In the minds of the people who controlled the money, there would be no connection to her specialty. She could imagine what an institution would think when they heard her say, "Hold your hats, gentlemen. You don't know me, but I've just discovered a brand-new organism that behaves unlike anything anyone has heard of before. I believe it has been killing fish in North Carolina for years, and I would like you to provide me with the funding to prove it."

Still, just to make sure, she made several inquiries and wrote as many proposals—only to be proved right.

And so she was feeling discouraged on a sunny September afternoon in 1991 when she took a break from a sea grass project she was working on at the National Marine Fisheries Service in Beaufort, North Carolina, and sat on the edge of a concrete tank, pondering her dilemma. Maybe it wouldn't hurt to approach the NMFS director, Bud Cross, and seek his counsel, she was thinking, and then she saw him walking by, as though conjured by mental powers.

Calling him over, she explained her problem. "Bud, I don't know where to go. I'm at a loss as to how to pursue this. Any ideas you have would be appreciated."

This was the first the director had heard about her discovery and he was enthralled. And having seen enough of Dr. Burkholder's sea grass work to know she was a careful, conscientious scientist, he thought she might really be on to something significant. "Let me make some calls," he said.

Within a week he phoned to say that he had arranged for her to submit an abstract late and make a presentation of her findings at an upcoming international conference on toxic algae, which would be attended by the biggest names in the field.

∼∽ ∽∼ ∼∽

There was another development during this period that put a shine on JoAnn Burkholder's life and generated almost as much excitement as her discovery: she fell in love.

A mandatory teaching load at the university included a course on the ecology of aquatic plants, and on the first day of class she was delivering a standard orientation lecture when she noticed a student who, rather than visibly listening to her, kept his eyes on the floor, as though preoccupied. He was a lanky fellow who wore a blue denim jacket and off-white jeans, but she couldn't tell what he looked like because his face was hidden by long, unruly blond hair. Throughout her talk she caught herself glancing at him, and on one of those occasions, at the precise moment she looked his way, his head swung up and she found herself impaled by a pair of penetrating green eyes set in a lean, uncommonly handsome face. Quickly she rolled her gaze on to the rest of the class, but her thoughts remained on him. Never before had she had such a strong and instantaneous physical attraction to a man.

Several minutes later, she stole another glance, and it was worse.

After class, she gave herself a private lecture: "Don't be ridiculous. He's one of your students." When that didn't work, she tried yelling at herself: "Are you crazy?" Which was totally useless. Finally she arrived at a truce: "Okay, you can have a horrible crush on this guy. But you must never let him or anybody else know how you feel."

Not once during the semester did she veer from that course. In fact, it could even be said that she overcompensated. In spite of efforts to appear natural, she was archly formal in his presence and more demanding on him than on her other students. In part, this was to keep her feelings in check, but it was also because he was given to make irreverent and flippant comments in class. Although there was some of that in her own personality, it was not a side she had ever exhibited to a professor, and she made a point of letting him know that she considered it inappropriate behavior.

It never occurred to her that her feelings might be reciprocated, until the course was over and he came to her office to pick up his grade. After lingering and making several awkward stabs at small talk, he came out with what was on his mind.

"I was thinking that after the holidays, when I get back from Christmas, maybe we could go out and have dinner together?"

His timing was bad. She was annoyed with him because by now she knew he was extremely bright and she felt he had done the absolute minimum amount of work it took to get his A. After hesitating, she replied, "Maybe." And then, to get him to hurry up and leave, she excused herself, going across the hall to an empty room until he was gone.

It took him three weeks to call. Over an Italian dinner, he admitted it had taken him that long to work up the nerve after her cool brush-off. She liked his candor. And the more she listened to him talk, the more she liked Mike Mallin.

It went slowly over the next two months as they got to know each other through dinner conversations and over drinks after a movie. Then it was "Thanks for the evening" and a quick kiss good night, because neither of them was inclined to rush into a serious relationship. Certain issues had to be worked through first, a big one being that she was a professor and he was a student, albeit working toward a Ph.D. at another university. Even though they were the same age and there was a good reason that he was behind her—after getting his master's, he had worked as an environmental scientist in industry for ten years—still, the "traditional male thing," as she would describe it, had to be dealt with.

Another brake was that JoAnn Burkholder was married. It had been a youthful mistake—a precipitate wedding back in 1977 to a fishery biologist she had met at Iowa State, whom she had left four years later —but even though they had been separated for almost ten years, neither spouse had bothered with the paperwork of a divorce.

On the other hand, there were strong affinities between her and Mike. They both enjoyed outdoor activities. He was familiar enough with the university scene that she could talk about it to him without having to explain the significance of departmental and publication pressures. And when he let her know, laughingly, that he had blown off her class one day to attend a Pink Floyd concert, she forgave him with the admission that she too loved rock music.

With sexual tension between them mounting palpably, they attended a Bruce Springsteen concert together. Afterward, they went back to his apartment. And rock-lovers notwithstanding, they would remember that night less for Springsteen's Raleigh stop on his "Tunnel of Love" tour than as the occasion when they first slept together.

They had seen each other steadily ever since, maintaining separate quarters but spending almost every weekend together and taking advantage of romantic getaways when they could. And so JoAnn Burkholder talked Mike Mallin into accompanying her to the Fifth International Conference on Toxic Marine Phytoplankton, with the promise that they would take a short vacation together when it was over.

The conference was held in a posh hotel on Goat Island, just off Newport, Rhode Island, in late October of 1991. After checking into a

Quality Inn on the mainland, which better fit their budget, they drove across the bridge to Goat Island to register.

As she stood in line awaiting her turn to sign in, Dr. Burkholder looked around, hoping to see someone she recognized. But the crowd in attendance was as strange to her as she was a stranger to it, and the case of nerves she'd been battling over the previous weeks returned in force. She believed wholeheartedly in her research. She had laboratory data, she had photographs, she had a historic record to support her thesis. There had even been another major fish kill on the Neuse River just before she'd left for Rhode Island, and it was one she'd managed to arrive at while it was still in progress. As she stood on a pier with death all around her—the toll was a billion menhaden—the importance of her research, if it could contribute to ending the carnage, had swept over her.

Nonetheless she was apprehensive. She did not feel she was a good public speaker, and she was absolutely convinced that nobody at the conference was going to believe her. In their eyes, she was nobody. And yet she planned to stand up in front of these seasoned veterans from around the world, people who had studied toxic dinoflagellates for decades, and tell them that she had found something they had missed.

To avoid having to answer questions, she and Mike skipped the preconference mixer, but she attended each of the talks the following day. *Damn,* she thought, when she realized the sessions were not concurrent. At other conferences where she had given presentations, the sessions were multiple—different talks went on at the same time—so there were rarely more than forty people in the audience. The format here meant that most of the 325 scientists who had registered for the conference could show up for her talk.

Her concern about how she would be received seemed self-indulgent to her, however, as she listened to the different speakers and realized that she was part of something critically important. For underlying every session was a common message: The world's oceans are experiencing an explosion of algae blooms such as toxic red tide, and there is reason to believe that this is an early warning of a massive ecological breakdown.

Papers were presented that discussed the possible causes for this burst of activity. Some attributed it to the oceans' becoming overloaded with nutrients poured into the seas and spread by human activity via sewage, industrial sources, and agricultural runoff pumped into coastal waters. Others saw it as an early by-product of global warming, involving shifts in coastal wind and ocean current patterns. Of particular interest to Burkholder was the claim made by several scientists that whatever the cause was, new types of toxins were appearing—previously unknown organisms and organisms once thought to be harmless.

Although Burkholder had been aware of the decline in the general health of the world's coastal waters, hearing so many knowledgeable people speak so specifically to the problem, and with such a strong sense of urgency, increased her sense of confidence as the time for her own presentation drew closer. What she was beginning to suspect was happening in North Carolina had been thrust into a larger perspective.

She took the stage just before lunch on the second day of the conference. She was faced with coordinating a microphone that wrapped around her neck, a pointer that shot a beam of light onto the screen behind her, and a slide advancer. She decided she would try to do without the microphone and simply project her voice. But no sooner had she put the microphone down and looked out over the crowded room when a small Japanese man leaped from his seat, grabbed the mike, and shoved it at her face. "You use microphone! Now!" he shouted.

It was such a startling and preposterous thing to have happen that she laughed out loud—straight into the microphone. So her laughter, booming across the room, was the first sound that the distinguished audience of over three hundred heard come out of Dr. JoAnn Burkholder's mouth.

Figuring she had nothing more to lose, she took a deep breath and plunged ahead.

"We've been hearing about some fascinating organisms that we have known about historically for some time that have been causing significant problems in terms of fish kills and human health." She paused to catch her breath. "But the story I'm about to tell you represents what I think will be a new chapter in toxic dinoflagellate research. In this talk I'll describe a previously unknown organism that is killing fish along the southeastern coast of the United States."

A nod to the moderator, and the lights in the room dimmed.

She began with a map slide of the Albemarle-Pamlico estuarine system, which, she pointed out, was the second largest in the United States, after Chesapeake Bay. This is where and how it started, she said, and she proceeded to review the grim beginnings of her discovery chronologically, from the unexplained fish deaths in the vet school at North Carolina State to the identification of what appeared to be an entirely new species of toxic dinoflagellate that was extraordinary not the least because fish seemed to bring out its killer instinct.

There was a murmur from the audience, as she knew there would be. The accepted wisdom was that algae produced poison as a metabolic by-product or to protect themselves against herbivorous fish. What she was saying contradicted everything they knew. But she backed up her claim about these toxic hit-and-run attacks with slides that did indeed appear to illustrate just such attacks.

When the lights came back up, the moderator cleared his throat as if he were having trouble swallowing something, which evoked chuckles from the audience. "Well, that was certainly fascinating. Are there any questions?"

It seemed to Dr. Burkholder that everyone in the room raised a hand, which did not come as a surprise. Nor did the doubts underlying most of the questions, because from the standpoint of toxic dinoflagellate orthodoxy, everything she had said about this dinoflagellate was wrong. The cyst she was showing pictures of was wrong—it looked just like the cyst of a totally different group of algae, and she knew they were probably thinking she had culture contamination and was just too inexperienced to know it. As for its response to fish, that too was wrong. No one in the room had ever heard of toxic algae that exhibited targeted behavior toward fish prey.

But to her amazement, the undercurrent of disbelief to the first round of questions quickly gave way to scientific curiosity.

In the end, it took her almost a half hour to get off the stage. Even then she was surrounded by people for another forty-five minutes. And they weren't just asking her questions; they were congratulating her. Several times she heard it said that she was the hit of the conference. Perhaps more important, people encouraged her to keep up the good work and assured her she should have little trouble receiving funding to further her research.

Afterward, she was ecstatic. As insecure as she had been going into the conference, she was flushed now with optimism that she had found a new and exciting direction for her work.

The rest of the time in Rhode Island was almost as exhilarating. A major storm blew in, bringing rain clouds and violent seas that washed away the edges of coastal highways. She loved it when nature expressed itself with drama, and grabbing Mike, she walked with him out onto the rocks at high tide, to embrace the storm's furious beauty. They returned drenched with rain and sea spray, and that evening they strolled arm in arm down the cobblestone streets of Newport, admiring the blue-water sailboats, and shared a romantic dinner.

Years later, JoAnn Burkholder would remember how, at that time, she thought life was great and would only get better. Then, with a short, sarcastic laugh, she would liken herself to Little Red Riding Hood skipping merrily toward the meeting with the wolf.

6

Although press coverage of the conference focused predictably on the established scientists in the field, word of Dr. Burkholder's talk got back to the *News & Observer,* Raleigh's daily newspaper, and a reporter was assigned to write a feature article on this new species of predator that was allegedly terrorizing North Carolina's coastal fisheries. It came out around Thanksgiving of 1991, and it was a long piece that reviewed the history of the discovery, included quotes from the principals, and was illustrated by a cartoon of the dinoflagellate launching itself from the river bottom like a fish-seeking missile.

As she read the article, JoAnn Burkholder found her eye drawn critically to the errors that typically accompanied scientific stories written hurriedly by beat reporters; but by the time she was finished, she'd decided it wasn't such a bad job after all. The writer got enough right, and besides, the way she was thinking, any exposure was beneficial, because it could stir up interest and perhaps generate research moneys.

What she did not expect was Dr. Edward Noga's "ballistic"—her word—reaction.

Over the preceding months, the two of them had been acting as coinvestigators, keeping each other informed about their respective experiments and discussing which granting agencies would be likely to fund their work. Her understanding of their relationship was that they were collaborators embarked on a common cause. But even though equality was something they had agreed upon in principle, there were times when she found Dr. Noga so hard to read she wasn't sure exactly what their relationship was. Part of this impression had to do with his personality, which was more reserved than hers. But on several occasions she had also found his behavior confusing.

As an example, she was anxious to obtain the financing that would allow her and Noga to get started with some new experiments with the dino, but it seemed that when she tried to get specific about funding leads, Dr. Noga's mind was elsewhere. "Go ahead if you want," he would say, "but I think it's a waste of time."

They were discouraging words, even though he didn't seem to have any better ideas.

Accompanying his apparent pessimism about raising money was a go-slow approach toward publicity. While she thought going public could lead to opportunities, he was sour on the idea of speaking out, though he never said why.

An inkling of what was happening came the evening she was invited to make a presentation to a group of environmental people in Raleigh. As she spoke about their ongoing efforts to understand this organism and how, in her lab, they had been perfecting techniques that would allow them to collect new information, she was surprised to see Dr. Noga sitting in the audience. Afterward, she went over to him and asked him how he thought it went. His answer struck her as odd. After muttering something about feeling perhaps she was overstating her role in the story, he said he wanted to set up a time when she could show him the new techniques she had referred to.

She'd looked at him curiously, because it had not gone unnoticed by her that while continually inquiring into what she was doing and how it was going, Dr. Noga kept his own research largely to himself.

Nevertheless, with a sense of team spirit, she agreed to a time; and several days later, accompanied by one of his technicians, Dr. Noga came over to her lab. After bringing out slides and microscope samples and giving them a demonstration of the protocols she was using, she said she had a meeting to go to, but they were free to stay as long as they wanted, just closing the door behind them when they left.

The next day when she came to work, she found a terse note in her mailbox from Dr. Noga, reminding her that *he* had found this organism, not her, and warning that she should respect the fact that *he* was the lead investigator in this research.

She didn't know what he was talking about. All along, whenever she spoke about the organism, she tried to be fair in crediting the role Dr. Noga and his assistant Stephen Smith had played in discovering it in their laboratory. But she had also stated—accurately—that she and her laboratory assistants had identified the dinoflagellate in the Pamlico Estuary, and connected it to prior fish kills.

That evening she showed the note to Mike. "Look what Noga wrote me. Can you believe it?"

"You know what you should do with this?" Mike advised. "Pretend it never existed. He was probably having a bad day. Forget about it."

From the time Dr. Burkholder first met Dr. Noga, there had been something undefinable about him that made her uneasy. Maybe it had something to do with his failure to make eye contact. She was also uncomfortable with the way he kept his office dark. On a beautiful, sunny day, his blinds would be drawn tight, as if he had a grudge against light. Now this nonsense.

But she knew Mike was right, and by the next morning she had cooled down and decided to take his advice.

Then the *News & Observer* article ran, and she received a second note from Dr. Noga, who, even though he was quoted extensively throughout, apparently was unhappy at having to share billing with her. He said he continued to resent the way she talked about this dinoflagellate. From now on, she was not to talk to the press without consulting him first. Any statements by her would go through him. Was that understood?

She didn't know whether to laugh or lose her temper. Was he serious?

Apparently he was. And while she had been willing to overlook Dr. Noga's intemperate first note, she was not about to let the second one pass without a strong response. She intended to set him straight before their professional relationship went any further.

She was trying to decide whether to write him a letter, call him on the phone, or set up a face-to-face, when something happened that suggested that the depths of Dr. Noga's anger toward her, and his possessiveness about this dinoflagellate research, bordered on the dangerously irrational.

A little more than a year earlier, having been awarded a small grant to study the effect of nutrient enrichment and sediment loading in a reservoir outside Raleigh, Dr. Burkholder had advertised in the newspaper for a laboratory assistant. It was a part-time position, and all she could afford to pay was seven dollars an hour, yet some forty people had answered the ad, and among them had been a higher-caliber candidate than she had a right to expect.

His name was Howard Glasgow, and he was a good-looking fellow in his early thirties, with a broad, boyish face, brown hair and eyes, and an easygoing, winning way that made him instantly likeable. He did not have a background in limnology, but he did have bachelor degrees in both chemistry and marine biology, as well as an expert knowledge of electronics. As he explained why someone with his abilities and background was unemployed at his age, she learned that after bailing his father's electronic service business out of bankruptcy and building it into the second-largest enterprise of its kind in the state— an achievement that earned him the Businessman of the Year Award in Durham and inclusion in *Who's Who Among U.S. Executives*—he had walked away from the job, literally, because of his father's extravagant spending habits. Figuring there was no better time than now to make a career change, he had decided to pursue his lifelong dream of becoming a scientist.

Glasgow was such a promising find that Dr. Burkholder had been reluctant to break the news that what he would be doing—fieldwork, collecting samples, and general lab duties—could probably be considered "science" only in a broad context. But she had been straight with him, and he had been straight with her, admitting that his intention was to go back to school for his doctorate, but if she hired him he would commit to a year.

Within a month after hiring Howard Glasgow, Burkholder realized her good fortune. One explanation was all it took for him to grasp a concept or a request. He had an intense, childlike curiosity to learn new things, he was incredibly detailed, and he was dependable. He also put in long hours without complaint until a job was done.

As soon as she could afford it, she had moved him to full time and raised his pay, and over the following months he had evolved into her right-hand partner in the lab. He helped develop research methodologies. He could implement chemical ideas. His knowledge of instrumentation design was invaluable. In time, whenever she needed something done or when something went wrong, she usually called on Howard Glasgow, her all-purpose troubleshooter.

Like everyone else who heard about the shape-changing dinoflagellate, Glasgow had been intrigued. He wanted to learn more, and even though they had yet to receive funding for research, he had told Dr. Burkholder that he would be willing to work with it on his own time, on evenings and weekends. So after her return from the meeting in Rhode Island, they discussed what experiments to begin with and decided that since the dino seemed to photosynthesize like a plant, he should initiate a series of challenges designed to determine the optimum light requirements for its growth and the effects of low light and darkness.

At this time, the toxic culture was still being grown in the environmental chamber at the vet school, where the dino was kept in aquariums fitted with filters and kept bubbling—it liked aeration—and the variables to be tested were added, followed by fish. The scope work and water chemistry analysis were then done in a lab in the basement of the building where Burkholder's office was located. So the day after the *N&O* article appeared, Howard Glasgow had gone over to the vet school to prepare for the light experiments, and the next morning he walked into Dr. Burkholder's office shaking his head.

"What's the matter?" she asked, and he told her that he had gone over to the environmental chamber the previous evening and Dr. Noga had thrown a tantrum. She watched him, waiting. And when he didn't continue, she said, "Yes?" meaning *Please explain yourself.*

Glasgow then told her that he had walked in on Dr. Noga, who was already there performing autopsies on fish from the dino tank, and for reasons Glasgow didn't understand, Noga had acted like a wild man, allowing blood to squirt on the walls, scales to fly, and fish heads fall to the floor. Then, when one of the forty-gallon tanks wasn't aerating properly, he had yanked out the filter boxes, and in the middle of reinstalling the filter lines he'd given up and stormed out. He didn't come back, didn't clean up after himself.

Glasgow was unaware of the notes that Dr. Noga had written to Burkholder, and she saw no point in informing him now.

"What did you do about it?"

Glasgow gave her a smile and shrugged. "I cleaned it up. Wiped the walls and floors with bleach."

"Thank you," she said.

Though she kept her feelings private, Burkholder was seething. It was one thing for Dr. Noga to express his resentment to her on paper, but to throw a fit that spread toxic material around a lab for one of her assistants to clean up was an irresponsible act.

She said nothing to Noga, because she felt sure he would just deny it. Now it seemed clear to JoAnn Burkholder what was wrong. It had apparently dawned on Dr. Noga that this organism had big-ticket potential, certainly in terms of prestige and possibly even money, and he was seeing her not as a collaborator but as a direct competitor.

More eager than ever to receive funding that would allow her to accelerate her research on the dino, Dr. Burkholder spent the next several months writing and submitting grant proposals. She went to the Department of the Navy, which had issued a request for proposals concerning water quality in estuaries, and asked for money to develop a program of mapping the regions and locations where fish kills occurred, then setting up a network of volunteer samplers. She went to the National Oceanic and Atmospheric Administration with a proposal to develop a molecular probe for this dinoflagellate. She submitted a long proposal to an environmental section of General Electric . . . and nothing came of any of her efforts.

It was frustrating—doubly so after the encouragement she had received at the Rhode Island conference.

Then she got a break.

In the late 1980s, the Environmental Protection Agency had recognized that North America's estuaries were troubled places. Fisheries were down; pollution was up. In an effort to save the nation's coastal ecosystems, the EPA had set up a federal-state funding program to study the causes of the degradation and consider means of stopping it. Ten million dollars had gone to North Carolina to determine the cause of an epidemic of recent fish kills in the Albemarle-Pamlico estuarine system, but to date, all the study had done was provoke a heated debate. Spokesmen for North Carolina's billion-dollar commercial fishing industry claimed the study was a waste of time. They said they already knew what the problem was and pointed to wastewater from farms, factories, and towns. The state agencies charged with protecting the environment took issue, saying there was not enough evidence to prove that was the source of the trouble.

When APES—the Albemarle-Pamlico Estuarine Study—was brought to Dr. Burkholder's attention, she contacted the appropriate people and informed them that she had collected data that could very well be pertinent to the study, because it implicated a causative agent in the fish kills. A presentation to a technical advisory committee to APES was followed by several other meetings, which led, in turn, to Drs. Burkholder and Noga receiving $40,000 in research moneys. It wasn't a lot, but it was a start.

In the classroom, Dr. Burkholder taught her science students that sometimes the best way she'd found to learn was by accident. "You may pride yourself on thinking you are objective," she would say, "but we all have our preconceptions. Even though we may try hard not to bring them with us when we approach the study of a system, we usually do."

"Scientists are often the worst," she would go on to say. "While most of them claim to be objective, in fact their biases frequently flow into their work and influence their findings."

"The only way you are going to make any exciting advances and really do great things in your field," she philosophized, "is to walk up to a system and just observe it for a while. Let *it* begin to tell *you* what's going on."

Over the next few months, that was just what Dr. Burkholder and Howard Glasgow did: subject the dinoflagellate to different environmental variables and watch what happened.

Most dinoflagellates were marine, but there were a few species that preferred a freshwater environment, so studies were set up to search for the limits of this dino's tolerance for salinity. They found it throve in salinity gradients that ranged from fresh water to salt water.

The same went for its response to temperature. As a rule, most dinoflagellates were fair-weather organisms, appearing in greater abundance in the warm summer months and tropical climates, but the inhibiting temperatures for this dino were extremely low.

As remarkable as its adaptability to environmental factors appeared to be, however, it was the research done to clarify its life cycle that brought the most mind-boggling results.

One of the first discoveries that surprised Glasgow and Burkholder was the dino's means of reproduction. While closely observing its behavior during the fish-killing stage, they found that at the same time it was feeding on fish particles, sexual activity was taking place. Cells simply divided, producing two gametes, sex cells analogous to sperm and egg; and while swimming about, these gametes fertilized each other, becoming larger forms of the original cell, which, unless fish were present, retreated to the bottom of the tank, secreted a new cyst about themselves, and went dormant.

Or so it seemed at first.

The way the researchers learned that the organism also had an amoeba stage was typical of the entire discovery process: it happened by accident.

In an effort to understand what effect the dinoflagellate had on blue crabs and whether or not it was the same as with fish, Dr. Burkholder had asked an undergraduate student doing an honors project in her lab to run some exposure experiments and keep data sheets. A week or so later, the student came racing up to her: "All the dinoflagellates are dying."

"Wait, wait, wait," Dr. Burkholder said. "What are you talking about?"

"My crabs aren't being affected because all the dinos are dying."

That didn't sound right, so Dr. Burkholder asked her student if she had brought any water samples along. Yes, she had, so Burkholder put them under the microscope. Sure enough, the dinos were dying; and a search around the rest of the sample revealed why. Somehow a tiny ciliate—a microfaunal predator—had got into the culture, probably hitchhiking on the back of a crab, and it apparently had an appetite for this particular dinoflagellate, because its gut was packed with dinos.

The discovery of a natural predator that fed on the dino in a manner similar to the way the dino fed on fish was a development as promising as it was exciting. But no sooner had Dr. Burkholder's thoughts turned to biological controls for this organism than she noticed something else in the water: huge—almost twenty times the size of the dino—

bizarre-looking, bloblike amoebae. They had spikelike extensions, as though protected by thorns, and looked foreboding, like monsters out of a horror movie.

She had heard about amoeboid dinoflagellates, but she did not at once make a connection. It wasn't long in coming, however. Several days later, the same undergraduate brought her a sample of water that had been collected several hours after the fish in the tank were killed by the dinos, and as she was scanning the specimen, suddenly Dr. Burkholder stopped on something that caused her to say, aloud, "Oh my God." It was the dinoflagellate in its sexual stage; but instead of feeding on fish flecks or dropping down into a cyst, before her eyes it was undergoing an amazing transformation. As though something larger and of a different shape was trying to break out of the cell wall from inside, first one spindly arm shot out, then another and another, and when the transformation was complete, she was staring at the bizarre, bristly amoeba she had seen several days earlier.

This was something entirely new, and a series of follow-up experiments to discover the controlling factors that determined when the dino encysted as opposed to evolving into an amoeba reminded her that they had only just begun to understand the many mystifications of this organism.

And if there were any doubts that it had aspects they could not predict, those doubts were promptly dispelled by a new set of experiments.

After finding the little ciliate that preyed on the dino, everybody in the lab was excited. The natural predator didn't seem to be bothered by the toxin and would eat the dino as it was killing fish. But while observing the dynamics for longer periods of time, Dr. Burkholder came to an even greater appreciation of the dino's survivability. Under the microscope one day, she was following a little ciliate that was energetically running around snacking on dinos, when the remaining dinos, as if they'd had enough of this harassment, turned on the ciliate, circling it like crows attacking an owl. Fascinated, she continued to watch as the ciliate tried to slink away. But it got only so far before one of the giant amoebae moved into the field and, extending its armlike pseudopodia, gobbled it up.

It was a stunning revelation. In the back of her mind, Burkholder had been thinking that maybe the amoeba would cannibalize its smaller life forms. Now she realized that that wasn't going to happen; for in its larger, more advanced stage, the dinoflagellate actually appeared to be acting in a somewhat protective role toward the smaller stages in the life cycle, coming to the rescue of its little brothers and sisters.

By this time Burkholder had identified almost fifteen different steps in the dinoflagellate's life cycle. What she still knew very little about, and felt was important to begin to understand better, was its toxin. What kind of poison, exactly, did it emit? What was its chemical structure? How, precisely, did it affect fish—that is: what did it do to a fish's nervous system, what to its flesh? And finally, did the toxin move up the food chain? In other words, would creatures who consumed fish hit by the toxin themselves be affected?

These were not questions Burkholder was trained to answer. But she knew who was. At the Rhode Island conference, she'd been given the name of Dan Baden, of the Rosensteil Institute at the University of Miami. He was reputed to be the world's expert in toxin analysis, an absolute ace at characterizing marine toxins. But she had also been told that he could be a difficult person to deal with, who was very busy and might not give her the time of day.

Late on a Friday afternoon in the spring of 1992, she dialed one of the numbers she had been given for Dan Baden. A man answered, and the conversation, as she recalls it, went like this:

"This is JoAnn Burkholder. I'd like to talk to Dan Baden."

"This is Baden," the man said.

Warned that he could be abrupt and she would have to make her case fast, she matched his terse tone. "I've been referred to you by a number of people. I have an alga that might be killing a lot of fish. It's a dinoflagellate. I really need some help in identifying the toxin. If you'd be interested in taking it on, that would be great. If you aren't, I'd appreciate your recommending someone who would. And if you're thinking this is a waste of your time, just hang up the phone."

There was a long pause. "Talk to me."

The short version of what followed was that he was interested. He said he had a window in his schedule at the beginning of July, and if she wanted to bring the toxin down, he would take a look at it for her.

On the evening of July 3, she and Howard Glasgow were driven to the airport by one of her graduate students. In addition to their luggage, they were carrying a five-gallon plastic jug with a screw-on top, which contained the culture in a highly toxic state. The organism required an unidentified substance from fresh fish to initiate toxin production, and in order for them to maintain a toxic culture it was necessary to feed it continually, exchanging dead tilapia with live tilapia quickly and constantly because the toxin literally ate away the fish tissue, producing "fish syrup" if the carcasses were not removed, ruining the water in the culture for use in characterizing the toxin. They kept up this exchange until just before boarding the plane, because they didn't want the dino to stop making toxin and retreat into a cyst state before they arrived in Miami.

Jostling was something else that made it go down. Sometimes just swirling it in a beaker was all it took for the dino to play dead. For that reason, they felt they could not send it cargo but had to hand-carry it, as gently as they could.

The grad student, convinced they'd never get away with it, made a wager they wouldn't get it on the plane. But they did, by way of a crazy scheme: they pasted on the jug a label that said "Betty's Mineral Water" and passed through security without a hitch, waving to the head-wagging student from the other side of the metal detector.

It was almost ten o'clock by the time they arrived in Miami, and because of the July Fourth holiday, they had a hard time finding a place to stay. Finally, around midnight, they located a Howard Johnson with a vacancy. Glasgow took one room, and Dr. Burkholder and the toxic dino checked into another.

She had no fish to feed it, so she could only hope for the best. Early the next morning, before breakfast, she and Glasgow found an aquarium shop and bought fish and a variety of accoutrements, including air stones and filters, because they intended to set up a complete culture for this ace dinoflagellate toxicologist.

They met with Baden around two o'clock that afternoon, and in person he was far less intimidating, was even charming. They carried the culture into his lab and set up a tank, and to give him a quick demonstration, they added some guppies and several mosquito fish in an effort to chum the dino into a feeding frenzy.

Nothing happened. The dinoflagellate wasn't interested. As they would later learn, these two fish were some of the least susceptible of all they would put to a test.

"These fish are not going to work," Burkholder said finally.

By now not only was Dan Baden watching the culture with a skeptical eye, he was appearing unsure of his visitors.

"Is there a place nearby where we can get some other fish?" Burkholder asked.

Baden seemed reluctant to volunteer the fact that there was an aquaculture facility across the road, but finally he did, and they rushed over and brought back several red drum.

And the fish started to go, exhibiting the first symptoms of neurotoxological stress. But the gleam in Burkholder and Glasgow's eyes faded when, just as suddenly, the fish began to recover. The dino had apparently had enough and was calling it quits.

Everyone sat there, and no one spoke.

Baden interrupted the silence. "Well, I don't know about this. . . ."

Thinking fast, Burkholder explained that the organism was cantankerous in culture. Sometimes it almost seemed that it sensed it was being studied and didn't like it. For no apparent reason, it would sometimes shut down, and it would take days or even weeks to bring it back up.

Before she had come, Baden had told Dr. Burkholder that he'd arrange for her to give a talk about the organism to a group of aquatic scientists at the Rosensteil Institute. Now he said to her, "Why don't we go through the slide presentation you intend to give, and you show me what's been happening."

The presentation was similar to what she'd given in Rhode Island, but there was no comparing the audience reactions.

"This is ridiculous," Baden said when she was finished. "I can't let you go through with this. No one will believe you. I don't believe you. I'm going to have to cancel the presentation."

Burkholder was mortified. All she wanted at that moment was to absent herself from the situation. So when Baden, seeing her dejection, apparently took pity and suggested she and Glasgow take the afternoon off and visit Sea World—he'd pay for their tickets—she almost welcomed the idea of retreating into the role of a mindless tourist.

She excused herself to go to the rest room, where she fought back tears of frustration. After splashing water on her face, she headed back to Baden's office. As she trudged down the hall, she saw Baden ahead of her, leaning against the doorjamb of one of his colleagues' offices, talking rapidly and excitedly, and she slowed her pace in order to better hear what he was saying.

"You wouldn't believe what I just saw—*I* almost don't believe it— but she may be on to something exciting."

The contrast between the way Baden had spoken to her moments earlier and how he was talking to his colleague now suggested that the doubts he'd expressed were a performance and he really was intrigued with this organism.

He hadn't heard her coming up behind him, so he didn't realize she'd overheard him until they were back in his office. There, she asked Glasgow to leave the room, and when she and Baden were alone, she confronted him.

"I came here to ask you to help with toxin analysis, not take what you damn well decide pleases you to take and do with it what you want," she said furiously. "If you're not interested in helping us, tell us so and we'll thank you for your time and leave."

A smile pursed the lips of the toxic-dinoflagellate ace. "Okay, okay," he said.

Reluctant to call the trip a bust, they stayed on for several more days, trying to raise the dino. But it simply would not cooperate, and in the end, all that came out of the trip to Florida was two free tickets to Sea World.

7

It was a time when she should have been leading a major research effort on an important discovery, with the best resources available to her, and circumstances were forcing her to take a seat-of-the-pants approach on a shoestring budget. A time when her fellow scientists were expressing skepticism and disbelief about her findings to her face, while behind her back they were trying to grab the dino research for themselves. In January 1992, she had been invited to speak at a seminar put on by the National Marine Fisheries Service in Charleston, South Carolina, which was trying to position itself as a national lab for testing and analyzing marine toxins. Three times over the course of the seminar, a top official made polite requests for a sample of her culture. In this business, cultures were gold, and when she just as politely turned him down, he assured her that one way or another, he would find a way to get it. "Just watch," he said.

This let her know it was also time to hurry up and publish. Even though there were aspects to this dinoflagellate that remained a mystery—in the lab, they had observed stages she was convinced were part of its life cycle, but she just didn't know where to fit them—she felt they knew enough to approach a reputable scientific journal. At the very least, publication would guarantee a certain amount of "idea ownership" and scientific credit for the work she'd done.

Sensible as this step was, the decision to take it was not an uncomplicated one; Burkholder felt that before she could go ahead and start writing, out of professional courtesy she needed to consult with Dr.

Noga. Relations between them remained strained. He had continued his pursuit of her research, making repeated orders for photographs, data, and summaries each time Howard Glasgow went over to the vet school.

When Glasgow relayed those requests to Dr. Burkholder, her reply was emphatic: "Absolutely not."

Only once had Dr. Noga shared the direction of his research with Burkholder, and then it was a slip of the tongue. The two of them had been asked to give a joint talk at a state fisheries meeting, which had obliged them to consult on who was going to say what.

When Noga said all he planned to divulge was that this dinoflagellate killed fish, Burkholder had responded, "For God's sake, Ed, you've got to do better than that."

"Well, I just don't want to let anything out yet."

"Like what?"

Noga cocked his head. "Like the fact that the toxin knocks the white blood cell count in fish down to twenty percent of normal."

Her eyes widened. So it wasn't just eating holes in fish; it was attacking the immune system. A drop to 20 percent was serious stuff.

"Yeah," Noga went on to say. "It's like fish AIDS."

"Ed, you've got to get that out," she said. But he had balked. And he never let her see his data and never mentioned it again, leaving her to wonder what else he had learned and was keeping to himself.

Her personal feelings toward Dr. Noga notwithstanding, when they had discussed publishing they talked about joint authorship. So Burkholder felt honor bound to discuss the idea with him again.

"It's March, Ed," she reminded him the next time they had an occasion to speak. "It's nine months after we first caught it in the act of killing fish. We've got to get out there and publish what we know before someone else beats us to it."

"I don't have enough yet," he replied. "I want to wait and do everything all at once."

"I don't think we can afford to wait much longer," she insisted. "I've talked to the press. You've talked to the press. Someone is going to scoop us if we don't write something."

When Noga was slow in answering, she rolled her eyes.

"Okay," she said, "how about we do it this way: I'll try and write something on its life cycle and fish kills, and you write about the fish pathology. Whoever gets a paper published first will be first author. I'll be second author on yours, you'll be second author on mine."

He looked at her. "Where would you send yours?"

"I'm thinking *Nature.*"

"Ooh, *Nature,*" he said, with a smirk that made her want to slap him.

Obviously he didn't think she stood a chance of being accepted in such a prestigious publication, and in fact she herself doubted she would make it into *Nature,* which had something like a 99 percent rejection rate. But she had decided to start at the top. She had considered *Science* but knew they were oriented toward the molecular and did very little on biology and ecology, while *Nature* seemed to have a more balanced acceptance of topics. Another part of her thinking was that *Nature* was published in Great Britain, and publication in an international journal would give her more of a worldwide readership, not to mention greater credibility.

For all JoAnn Burkholder's insecurities about public speaking, when she sat down to write, she was in absolute command. (In undergraduate school at Iowa State, she had been exempted from taking a scientific-writing course simply by showing her professor a science paper she'd written as a high-school junior.) She wrote this paper with *Nature*'s format in mind. Feature articles were generally long pieces that restated what was known about an accepted subject, then presented the latest information. A section headed "Letters" published shorter scientifically refereed papers, to let the world know about something previously unknown. So she decided to pitch her paper as a "letter."

Nature usually let submitters know within two weeks whether a manuscript was accepted or rejected, but a month went by and Burkholder heard nothing. Cautiously she allowed herself to be hopeful, interpreting the fact that it was not returned immediately to mean it had been sent out for peer review. Six weeks passed—to her it seemed much longer—before she received a letter of acceptance.

The next morning, she bounced into the basement lab where Howard Glasgow was working, sailed the letter across his desk and held out her hand, palm up. When he'd read it, he let out a whoop and slapped her hand with his own.

Following closely in the mail was a copy of her manuscript with editorial comments and copies of the peer reviews. Prior to acceptance, *Nature* had sent the paper out to two recognized leaders in the field. Eagerly she read their opinions, and key phrases jumped out at her from the first review: "strongly recommend publication . . . findings represent major new discovery that may be of substantial general relevance . . . research of high quality and carefully executed and presented."

The second was even better: "recommend acceptance . . . organism is extraordinary in behavior and represents a new dimension in the world of toxic phytoplankton . . . readers will be fascinated . . . author calls it 'phantom-like,' although I think a more apt simile can be found in gothic literature."

She was overjoyed. But then she remembered that she would have to inform the second author, and even though she and Dr. Noga had agreed verbally to a you-try/I'll-try deal, she knew he was going to take it hard.

A bit sneakily, she got Noga to her office under the pretext that he would receive research data he had been after for a long time. She didn't know how else to do it. He was sitting on the opposite side of her desk when she broke the news, and he froze, so affected that he was virtually unable to form a sentence.

"Why . . . why didn't you . . . why didn't you tell me?"

"I'm telling you now," she said. "And I've been telling you for months we needed to write a paper. But you wouldn't do it. So I did. And if you'll remember, this is exactly the way we decided to do it. I would write a paper on the life cycle and fish kills."

Dr. Noga seemed to be having comprehension difficulties as well as verbal ones.

"I just don't understand. How could you do this?"

Losing patience, she thought it disingenuous of him to try to induce guilt. "Ed," she said, "make comments before it goes into galleys, and I'll incorporate them. As for publication of the paper, it's a done deal."

When Noga had stiffly exited her office, Dr. Burkholder thought, *I'll bet he's going to try to contact* Nature *and say he should be the first author.* She even suspected that he might call the editors and accuse her of stealing his research. It was something she knew he could never prove, because she had made absolutely sure that everything in the paper had come from her lab, but still it could foul things up with the editors.

That didn't happen. With very minor revisions, the paper appeared in the July 30, 1992, issue of *Nature* under the heading "New 'phantom' dinoflagellate is the causative agent of major estuarine fish kills." Concisely written and illustrated with eight photographs, it covered the period of time from the dino's discovery as a "culture contaminant" to the confirmation of its lethality to both finfish and shellfish. But it didn't end there. When she'd written the paper, Burkholder had wanted to be sure that readers did not consider this a parochial problem, endemic only to North Carolina's estuaries. Few algae were found in just one place in the world. In fact, many of them were cosmopolitan, and she suspected them as the cause of unexplained fish kills in other parts of the world. All of which led to the last sentence in the paper: "We predict that with well-timed sampling, this alga will be discovered at the scene of many fish kills in shallow, turbid, eutrophic coastal waters extending to geographic regions well beyond the Pamlico and the Neuse."

The response was almost immediate, and it was overwhelming. For the next several weeks the phone rang constantly, both at her office and at home. Newspapers from around the country, from around the world, wanted to interview her. They also wanted pictures of the organism, and she spent more than four hundred dollars over the next two months duplicating slides and photographs.

The *Washington Post* found the story so spectacular, so just-this-side-of-sensational, that it couldn't help itself: it had to make sport of the discovery while reporting it as news. The headline on the piece it ran read: "Look! In the Water! It's a Plant, It's an Animal, It's a Toxic Creature!"

Coverage in the *San Francisco Examiner* included a quote from a local phytoplankton ecologist, which would be widely circulated for its colorful aptness. According to him, the bizarre phenomenon of an aggressive amoeba-vegetable actually hunting down fish to eat was "analogous to grass feeding on sheep."

Not about to be left behind, North Carolina State featured Dr.

Burkholder in the campus paper and the alumni magazine, showcasing the pioneering work being done by one of its best and brightest faculty members.

Of course, she also received a fair share of inquiries that fell into the crackpot category: chemical companies, for example, that claimed they had just the right formula for getting rid of this dinoflagellate, as though it were a prehistoric roach and they had been doing research in anticipation of the need for just this sort of extermination. She also had more offers than she cared to count from scientists around the world who said they were willing to help her if she would kindly send them samples of her culture.

This was Burkholder's first experience of the bedlam that could accompany public notoriety, and her recall of the next three months is a blur. She remembers certain radio and newspaper interviews and being bombarded with questions. She also remembers calls and letters of congratulation, and having to deal with errata in the *Nature* paper. It had been her understanding that Stephen Smith, Dr. Noga's laboratory assistant, would be put on Noga's publications but not hers, until Smith contacted her to say that Dr. Noga had promised him that he would be included as an author on whatever came out first about the dinoflagellate. A funding agency that had given her a grant also complained because it had not been mentioned in her acknowledgments, even though the work it had supported was completely unrelated.

From time to time during this period—usually on quiet hikes that she took to hold on to her sanity—Burkholder would reflect on the impediments that had blocked the road to recognition. And again and again she found herself thinking about the politics of science and how unhelpful her graduate-school training had been in preparing her to play the game.

Nowhere was this more evident than in the reaction to her paper from the toxic-dinoflagellate crowd. Perhaps they considered her a trespasser on their turf. Perhaps they didn't know what to make of her story and thought she might be making it up. She could only speculate. But except for a few, they mostly expressed skepticism or ignored her findings.

When at last Dr. Burkholder was able to return her attention to her work, she was particularly focused on divorcing her research effort from Dr. Noga's operation. The situation at the vet school had become intolerable. On several occasions, Howard Glasgow reported, Dr. Noga had come down to the environmental chamber and actually demanded that he be allowed to sample Glasgow's experiments.

Primed for a confrontation, she went to speak to Noga about his inappropriate demands. It was a "creepy" session. She would talk, he would talk; and when she replied, he would stare into space. A transcript of their conversation would not have revealed anything out of the ordinary, but to her those lapses meant everything. She not only felt he wasn't listening, she sensed a bitterness so extreme that he found it difficult to sit in the same room with her. Which was why he kept drifting off.

The next time Howard Glasgow said he'd been approached by Dr. Noga, Burkholder drew the line. "Scramble the labels," she said. "We've got to get out of there."

Following her instructions that day, Glasgow removed the labels from the flasks in the environmental chamber and mislabeled them deliberately, according to a code.

Finding a separate facility where they could continue growing the dino took time. She would locate an acceptable space, only to encounter resistance from other researchers in the building: *You're working with a fish-killing dinoflagellate? Well, we don't want it anywhere near us. I don't know what it does, but I'm not going to take any chances.*

Meanwhile Glasgow would show up at the environmental chamber and find his mislabeled flasks removed from their shelves and shoved into a corner to create space for Noga's flasks.

We have to find somewhere else to go or else we'll be out of business, Burkholder thought.

It was late September before an alternative facility was found, a general lab in the Department of Atmospheric Sciences building. It was understood that this was only a provisional arrangement until something permanent could be set up, but at least it was a place she could call their own.

It was not the end of her troubles with Dr. Noga, however.

Not one month later, Howard Glasgow came to her with a story that made her so furious with Noga, "if I'd had a gun I'd have wanted to shoot him."

She was in her office when Glasgow called from the new lab. He said he was just checking to see if she was in, because he had something he wanted to talk about. "Come on up," she told him.

He arrived, his usual smiling self. But when he started off by saying, "Man, something really bad happened to me last night," she knew he was deadly serious.

"What's going on?" she asked.

With a slight embarrassment, he admitted that, on the side, he had been helping Dr. Noga at the vet school.

When she started to sputter, he explained. "I didn't want to tell you about it because I knew how you'd feel. I was doing it just to try and keep the peace between you two."

Eyes narrowed, Burkholder shook her head.

"So the other day, Dr. Noga called and asked me if I would do him a favor and change the fish in his aquaria."

"Son of a bitch. He asked you to do this? He's only three floors upstairs, and you're twenty minutes away. What is he—too lazy to do it himself?"

Glasgow waited for her to calm down.

After a moment, she said, "Well, that's bad enough. What other good news do you have to tell me?"

A look she'd never seen before crossed his face. "I ended up crawling out of there."

Then he told her how, late the previous evening, he had gone over to the chamber and found it a mess. How some 500 gallons of toxic material had been allowed to evaporate, and toxic slime coated everything. It was dripping from the ceiling; the empty aquariums were sticky with it. He had been in there about twenty minutes, cleaning up, when he found himself getting short of breath. Then he lost feeling in his legs, his coordination went, and the next thing he knew, he was on his hands and knees, vomiting. He said he could not think clearly, but somehow he managed to crawl out of the chamber.

"Damn!" Burkholder exclaimed. "What happened after that? What did you do?"

Glasgow said he sat on the floor outside the chamber, leaning against a wall until he regained his faculties. Then, after about an hour, he searched out a respirator and went back into the room to clean up after himself.

Words could not express Burkholder's feelings at that moment. Dr. Noga was a highly trained scientist in clinical/medical procedure. He had to be aware of the working conditions he had established. And he had asked *her* assistant to do his dirty work.

It was at this point she was glad she wasn't armed.

"You are never to go back there again," she said. "Period. No matter what happens. Do you understand?"

Glasgow nodded slowly.

"And I will call Noga and tell him, if you don't want to."

Although Glasgow gave no indication that he wasn't up to making the call, he looked uncomfortable at the prospect, so Burkholder went ahead and made it herself.

"Howard will not be helping you in your laboratory anymore," she said when she had Dr. Noga on the line. "And I would suggest you clean that filthy pigpen so nobody else gets hurt."

With disgust, she heard him plead innocence.

"JoAnn, I don't know what you're talking about."

"I don't damn doubt it," she snapped. "You're never down there yourself. You send everybody else down into that death chamber."

"I had no idea," he muttered before she slammed the phone down.

Actually, this was not the first episode Howard Glasgow had experienced in the environmental chamber. About a year earlier, just after Dr. Burkholder returned from Rhode Island, he had gone in to clean the chamber—which had not been used in some time—and to prepare the aquariums for some light experiments. The hose had a fine-spray nozzle on it, and kneeling next to a drain in the floor, he had been rinsing a ten-gallon aquarium when a strong mist blew up in his face. He didn't think anything of it at the time, not until he stood up and reached for another tank and realized that his thoughts were racing far ahead of his movements, which seemed to be proceeding in slow motion. Sensing trouble, he started to walk out of the chamber, and it seemed to take forever for him to pick up one foot and put down the other. He later likened the sensation to moon walking. Once he got outside, it took about fifteen minutes before everything returned to normal, and then he wrapped up the hose and got out of there.

Afterward, he didn't make much of what had happened. He thought maybe he'd had an acute allergic reaction to the atomization of the detergent and bleach he'd been using. When he told Burkholder about the incident, he described it as more like a drug high than a frightening experience. Grinning, he had shrugged his shoulders as if to say, Have fun figuring that one out.

Few researchers took lab safety for their staff as seriously as JoAnn Burkholder. It went back to a bad experience she'd had as a junior in college, when she received an acute short-term exposure to acetone and hydrochloric acid fumes while working in what was later discovered to have been a defective fume hood. As a result of the scar tissue in her lungs, she'd had difficulty climbing stairs without wheezing for months afterward; one night, she woke up hemorrhaging blood into a pillow.

At the very beginning of her present research, when it looked as if they were going to continue to work with a toxic algae, she had made extensive inquiries into whether or not there were safety precautions that should be taken. She had called the top people in the field for guidance, and the experts had suggested only that she follow "normal" precautions. This amounted to an advisory to wear gloves and a lab coat. A toxic dinoflagellate specialist with the Food and Drug Administration had even informed her that his toxic dinoflagellates were grown in large open bubbling vats, and no one had experienced health problems.

Everyone acknowledged that human susceptibility to poisons produced by dinoflagellates was associated mostly with consumption of them. Again and again Burkholder was told they could be nuisances, but they were not public health hazards.

Given the information she had at her disposal, when she considered Howard Glasgow's second episode, Burkholder applied conventional explanations: *Maybe it was simply a concentration of noxious fumes.* She ended up reasoning: *If someone were to walk into a room with a similar quantity of evaporated vinegar, perhaps a similar reaction would occur.* She continued to think along those lines for a little more than a month —until she had her own "episode."

Over December and into January of 1993, she spent more time than usual in the temporary lab, helping Howard Glasgow keep the cultures hot. On average, she was there an hour in the morning and an hour in the evening, exchanging dead fish for live fish. It was in the course of preparing an experiment on a Saturday in late January that she picked up a beaker filled with an extremely toxic batch of culture and began to pour it gently into another flask. The slightest jostling could shut the dino down, and she was trying to make sure she was adding just the right amount of volume. Holding the containers directly in front of her face, she was very careful as she poured. . . .

She can't remember very much of what happened after that. Just a couple of strong visuals and a series of reactions she would find strange only in retrospect.

She recalls, for instance, that her eyes began to burn, so she raised a gloved hand that was dripping with toxic water to rub them. She recalls that though her motions were becoming slower and slower, she felt compelled . . . to . . . keep . . . going.

She also remembers getting irrationally angry at Mike—who was in the lab conducting a joint experiment with her—when he accidentally dropped a chem wipe on the floor. As she shouted at him about his clumsiness, she was aware that she was overreacting, and even as she heard herself lose her temper, she was able to stand apart from herself and think, *What are you doing?*

The narcotic effect seemed to wear off fairly quickly. But by the time she had locked up the lab and was standing at the elevator, she suspected her troubles had only just begun. She felt awful. She was having trouble breathing, and she was suffering from severe stomach cramps. Mike drove her home and told her to stay there until she was better, but Sunday she still felt bad, and Monday was even worse.

All this while, she was thinking flu, because the symptoms were somewhat similar. But after Tuesday, she realized that something else was going on.

Teaching a class that morning, she misspoke and her thoughts were confused. It was as if her short-term memory were short-circuiting. Mercifully, her long-term memory and notes got her through the lecture, but she was unable to answer some very basic questions asked by her students.

After class, she sat in her office, dazed and disoriented. *What is going on?* she wondered. She walked over to her computer, but by the time she got there she'd forgotten what she was going to do. She sat down, booted up a program, looked at the screen, and tried typing a sentence —but the letters seemed to float, and she couldn't rein them into sentences. She returned to her desk and found herself holding something she did not remember picking up. The phone rang and she answered it. But partway into the conversation, she realized that she couldn't remember what had been said at the beginning, so nothing now made sense. "I'm sorry. I have to go," she mumbled, and hung up.

She did not take any more phone calls. She headed home, and of that drive, all she remembers is looking at the speedometer and realizing that she was going eighty-two miles per hour down streets whose speed limit was forty-five.

Somehow she made it to her apartment safely; and after locking the door, she did not emerge for almost a week. She couldn't focus, couldn't concentrate, couldn't think, couldn't remember; everything was wrong, and she didn't know why. When it wasn't frustrating, it was terrifying, and she kept asking herself, *Is it always going to be like this?*

She tried to call Mike, knowing he had no idea what she was going through, but she couldn't remember his number. So she looked it up and started to dial but could only manage to remember one number at a time. By the time she was finished dialing, she hung up because she didn't know what to say to him and was afraid she'd be unable to retain what he said to her.

There were moments of lucidity. In the brief periods when she would snap out of it and regain her cognitive faculties, she was reminded of descriptions she'd read of Alzheimer's disease: windows on reality would open only long enough for sufferers to realize something was happening to them that they could not control, and then the windows would turn into doors and slowly close.

Although her recall of that week is incomplete, she remembers two images that kept recurring during her lucid moments. One was the vision of her gloved hand, dripping with toxic water, coming slowly up to rub her burning eyes. No one *in her right mind* would have done such a thing. At no other time in twenty-two years of laboratory work had she acted so carelessly. It was almost as if she'd lost her fright response, and taking its place was a compulsive keep-going-with-what-you're-doing behavior.

The other image was of a fish swimming dreamily in a water column as though trapped inside a bubble, unaware that the very cause of its euphoria was killing it, not realizing that in order for it to save itself, it had only to swim away. She found herself identifying with the fish, making an analogy between what it was going through and what she was experiencing . . . and then she'd drift back into a haze.

Toward the end of the week, she felt as though a spell that had been cast on her was weakening. To test her mind, she sat in front of the TV, watching the news, and when it switched to a commercial she would try to write down what she had just heard. At first she was unable to repeat even the last sentence of what the anchorperson said, and she threw her pencil across the room, lay back on the couch, and started to cry. But gradually and with practice her memory improved.

It wasn't until she was able to read and remember, to trust herself to follow someone's line of reasoning in a conversation, that she gave herself permission to resume her routines. She wasn't all the way back yet. She still didn't feel right physically, and there were intellectual deficits that persisted, most noticeably a lack of sharp, clear thinking. But she was driven to get back to work by the need to make sure nobody else in her lab went through what she had. Because when she put it together now, she knew that this dinoflagellate could no longer be considered a threat merely to fish.

8

Convinced that she'd been hit by an aerosol accumulation of concentrated aquarium cultures—which meant that the effects of the dinoflagellate's toxin could be transmitted through the air—JoAnn Burkholder confronted Howard Glasgow immediately upon her return to work. She wanted to know if he'd had a similar experience but kept it from her.

He swore he had not. Since they'd moved to their temporary quarters, he *had* noticed a slight shortness of breath when he entered the room, as well as a periodic eye-burning sensation. But that was it.

She had no reason to suspect he was holding back, nor did she believe that what she'd suffered was an idiosyncratic reaction.

"Okay, Howard, I want you to drop what you're doing and shut down all but one toxic aquarium. Let all the other cultures go dormant —I don't care. I don't want anybody working around this alga until we know how to incorporate better protective measures to make sure people will be safe working with it."

Burkholder began a round of campus calls. Having already made inquiries among researchers in the toxic-dino field and been assured there was nothing to worry about, she looked for guidance within her university, from people who could talk to her about safety issues as they pertained to biohazard environments.

She received an education. She learned that North Carolina State University had strong policies and programs to deal with radiation detection. And that the university had tightened up its chemical safety protocols after the Environmental Protection Agency had identified an area near the football field—where researchers in the 1960s and 1970s had buried their chemical waste—for placement on its Superfund list. As for biosafety, however, there was next to nothing in place. What guidelines there were came primarily from the National Institutes of Health's Centers for Disease Control, which associated safety recommendations with specific hazards, such as genetic engineering. And it went without saying that those guidelines were not written with this dinoflagellate in mind.

Indeed, the attitude of the university toward biosafety programs was summarized for her in a conversation with a professor who served on the faculty committee that advised on chemical safety issues. When she asked him what kind of biohazard controls he would recommend, he warned, "Keep up with these kinds of questions and you're asking for trouble. You'll have to deal with all sorts of regulations. You may even end up having your lab investigated."

His advice to her? "Keep your mouth shut and do it the way you want it done."

It was not a singular position that he was taking. For a lot of researchers, mention safety and immediately it was: *Oh, God, more rules and regs. Why don't they go away and leave us alone so we can do our work?* Academic investigators didn't like to have to deal with safety issues because in an academic setting, projects came first. You did what was necessary to be a successful investigator. Fame and fortune came from progress, not compliance. The issue for some was not safety but academic freedom.

Although it was clear to her that there was a hole in the university's safety program regarding biohazards, Burkholder continued to make inquiries. This was how she learned about a new Environmental Health and Safety Center, apparently created in response to a series of articles in the local newspaper exposing the unregulated research environment on campus. Dave Rainer was the center's director, and while he had come to his position after establishing a tough reputation in industry, his initial response when she described her situation was typically bureaucratic.

How do you know for sure that something in your lab is the cause of your problems?

"Because I was fine when I walked into the lab, and when I walked out I was not. I was really sick."

Did you file an accident report?

"No, because it wasn't like an accident where you fall and break a leg. Something strange happened to me when I was working under routine circumstances. I went home and hid."

You should have filed an accident report.

"Look, I had trouble forming a sentence. I wasn't thinking about normal accident protocol. This is my form of filing an accident report."

Perhaps you had a touch of flu.

"I know what flu is, damn it, and you don't suffer from short-term memory loss with flu."

Oh, short-term memory loss? That happens to me every night after dinner.

"What are you saying? I'm making this up?"

Though Rainer's skepticism infuriated her, she understood its basis. There was nothing in the literature about the toxicity of this organism except what she had written, and that referred only to its effects on fish. An entirely separate laboratory on campus—Dr. Noga's—had worked with this dinoflagellate for at least as long as she had, and as far as anyone knew, no one at the vet school had complained. So there was nothing other than her claim that tied this organism to an adverse human health effect.

Burkholder didn't know what to say. She was certain they were dealing with a pathogen that had the ability to volatilize, yet she couldn't prove it.

At this point, the director softened and told her the ideal situation would be for her to have her own facility, so that she could take whatever precautions she felt were necessary. But that was a matter to be taken up with the university. In the meantime, he suggested, she might want to try using face-mask respirators, and he handed her a pamphlet.

In the course of seeking advice from the "specialists," she had asked one prominent researcher if he recommended wearing masks in the lab, and he'd laughed: "Yeah, sure, a Halloween mask." He'd also said, "Dinoflagellates do not produce neurotoxic aerosols."

Left to her own resources, Burkholder decided to order full-face mask respirators for everyone who would be working around toxic cultures. Selecting the proper ones posed a slight problem, since the toxin they were trying to guard against had yet to be identified. Then she remembered that early on at the vet school, they had conducted a very simple experiment using carbon filters in fish tanks and found that they provided an effective barrier against water loaded with toxic dinoflagellate culture. Knowing that charcoal screened out organic molecules, she figured an equivalent filter would do the same for a respirator.

But even as she took this step, Burkholder thought of it as an interim measure, because she knew Dave Rainer had been right when he suggested that the ideal was an independent research facility of her own. She had already decided to settle for nothing less.

The sequence of events that converted that notion to a reality began with a departmental meeting at which she explained her discontent with their working conditions. "We're doing very important research from the perspective of understanding what has been going on in North Carolina's estuaries," she said. "But in terms of facilities, it's ridiculous. Nobody wants us around. So we've had to work in borrowed facilities for a few months at a time. We can't even clean our aquaria without physically taking tanks that are contaminated with toxic spores, putting them into a station wagon, and transporting them across campus to a place where we can hose them down. It takes us at least four or five times longer than it should to do the research, not to mention that we're lugging a toxic organism around in state vehicles that will be used by other people after us. We're as careful as we can be, but this does pose a hazard. I shouldn't have to operate like this, and I need your help."

The logic of her appeal was compelling. This meeting was followed by one with the dean of research, at which she expanded her argument to include the fact that North Carolina State had gained a lot of positive publicity in the national press from her research and stood to benefit even more if she continued to be successful. But for that to happen, she needed a proper laboratory.

When the College of Agriculture and Life Sciences informed Dr. Burkholder that she was being awarded $30,000 for the construction of an independent research facility, she was thrilled. But she was also overwhelmed by the thought of the time and energy that would be required for her to move ahead with her research and at the same time oversee a building project. She estimated that if she turned her attention to setting up a new facility, it could be as much as a year before they were back in the dino business.

"So leave it up to me," Howard Glasgow said when she unburdened herself to him. "I know my way around construction. I have a certificate in electrical engineering. I'll take it on."

Although Glasgow had committed himself to one year when he went to work for Dr. Burkholder as a seven-dollar-an-hour part-time lab assistant, three years had passed, and she now considered him a full-time associate. Indeed, the research methodology that had evolved between them was decidedly collaborative. The first step in the process usually took place in Burkholder's office, where they would conduct brainstorming sessions about the direction their investigations should take. Then they would design experiments, contriving conditions whereby observations could be made and data collected. Sometimes at this stage they would use a blackboard, inscribing select variables and controls, techniques for setting up different treatments, containers to be used and equipment required. Occasionally they would pull in other technicians and graduate students, taking them out for pizza and beer and soliciting their ideas, some of which proved very helpful. Finally Glasgow would put together a proposal, run it by Burkholder, and after more critiquing and some fine-tuning, it was often: "Okay, let's try it."

Howard Glasgow had become a terrific partner, and she thought they made a great team. So when he offered to take over the search for an independent facility, with as much appreciation as relief she told him, "It's all yours."

A series of phone calls led Glasgow to a zoology professor who was doing parasitic studies in a special modular unit—a trailer—located in an off-campus area that had been set up for researchers doing nasty, pathogenic work. From their discussions, Glasgow learned everything he needed to know about arranging for a comparable facility in the same location. After he'd set in motion a request to the university's department of campus planning for space in the "research annex," as it was called, Glasgow contacted the local distributor for the outfit that had built the zoologist's unit, to discuss the design and specifications of the facility.

The first question that came up was what safety classification he wanted to adhere to. He said he didn't know what was best, because their efforts to determine specific guidelines for dealing with toxic algae had been fruitless. Asked about the character of the hazard, he replied that the toxic features of the organism they were working with had not been identified, but they suspected there was some kind of aerosolization. He was told a general Level 2 biohazard facility should be sufficient. It was the next step up from standard microbial precautions and just below a Level 3 facility, which was recommended for experiments involving potentially lethal agents.

Using computer-aided design drawings, which he faxed to the contractor, Glasgow revised the specifications on the modular unit over the month of February and into March. He wanted the facility to be divided into two separate zones—a "hot" room, where the toxic organism would be grown and studied, and a "cold" room, a safe area where other activities could take place. He wanted the ventilation system to circulate fresh air from the cold room into the hot room, where it would be expelled by powerful ceiling exhaust fans that would provide a minimum of twelve air exchanges per hour. And finally, anticipating how much time he would be spending at the facility, he added a bathroom and a small office on the cold side.

As the plans neared completion, the representative for the contractor said they needed a name for the unit and asked what exactly it was going to be used for. When Glasgow told him it was for a toxic dinoflagellate study, the rep, who personified the term "good old boy," replied, "What in hell is that?" Glasgow explained, and the man joked, "Well, that's a mouthful for this country boy. How 'bout we keep it simple and go with 'fish laboratory'?"

Before the actual order for the fish laboratory could be placed, the blueprints and designs had to pass the university's review process. To save time, Glasgow personally delivered copies to the nine different departments whose approval was required, and when the signatures had been obtained, he gave the green light to the contractor, which in turn gave him a delivery date—only four months from the project's inception.

The day the trailer arrived, Glasgow was down on the coast, finishing work on a sea grass project, and driving back to his home in Durham, he decided to detour by for a look. To get there, he followed a campus road that ducked under a parkway, passed a nineteenth-century "poor people's" cemetery, and dead-ended in an area that had been cleared of pines and scraped flat. It was around five in the afternoon and raining torrents. Skid marks in the mud explained why the unit wasn't parked on its designated site but was sitting in a nearby field. Curious to see how it looked on the inside, he parked as close as he could and ran through the rain, hopping up through one of the doors.

As he stood in the cold room, glancing around, the first thing he noticed was that one of the overhead light fixtures was full of water. *What the hell?* he thought. He had specifically requested that special gaskets be installed around the lights, because he knew the work involved highly humid conditions.

After looking around some more, he went into the hot room; the overhead light fixture there was full of water too.

Glasgow knew that with any construction project, seldom was everything perfect to start with. Nevertheless it was disturbing to find a major problem with something as fundamental as a leaky roof.

Several weeks after the modular laboratory had been moved to its proper location, he met with the contractor's representative to discuss deficiencies upon receipt. He was assured they would fix the leak, along with some water damage to the floor; and they promised to complete the water hookups within a couple of days.

Glasgow had to make another trip to the coast, so it was Monday of the following week before he was able to check on the repairs. Opening the front door this time, he unleashed a flood of water that stood two inches deep throughout the trailer. When it had drained, he went in and investigated the source of the problem. A PVC pipe under one of the sinks had popped off under pressure and water had run continuously for the entire weekend, filling both the hot and cold rooms, buckling the floors, dripping through the seams, and waterlogging the insulation.

It was an ominous start to a project intended to improve safety, and there were additional problems and more delays before Glasgow could start moving aquariums over from the old lab. It was June before the new facility was up and running. Glasgow was so busy playing catch-up with his lab work that he didn't have time to worry about whether or not the facility they had discussed at length in the designing phase was the one he'd received.

～～ ～～ ～～

Despite the glowing endorsement provided by *Nature* a year earlier, Dr. Burkholder was still battling for credibility within the scientific community. Scientists don't like taking someone else's word for things —they want to see for themselves—and this organism was especially difficult to isolate. A technician scrutinizing a sample on a microscope slide crowded with thousands of cells of dozens of phytoplankton species was going to have a hard time differentiating the toxic from the nontoxic strains, for many looked deceptively alike. If you didn't know what you were looking for, and if you were dealing with a microscopic organism that took on almost twenty different disguises, it was easily missed.

For this reason, Burkholder conducted a series of seminars for university people and state agencies in several Atlantic-coast states, showing them how to find the organism under the microscope. It was also why she offered to fly the esteemed dinoflagellate expert Dr. Karen Steidinger from Florida to Raleigh for a demonstration. She'd been sending Dr. Steidinger live cultures for almost two years, but inexplicably the technicians in Steidinger's lab had reported that they were unable to observe the dino doing what Dr. Burkholder said it did.

There could have been any number of reasons. One of the problems with working with this dinoflagellate was that it preferred and gravitated toward a regime that had been established in the estuary, which was not easy to reproduce in a laboratory setting. That was true of a lot of dinoflagellates: they responded to indirect cues, little triggers in the medium, which were tricky to duplicate.

They must be doing something wrong, Dr. Burkholder thought. *We'll have to get her up here and show it to her.* She felt it would be worth paying Dr. Steidinger's travel expenses, because if someone of that stature corroborated her work, it would go a long way toward silencing the skeptics once and for all.

The visit took place in late spring of 1993, and it started poorly. The first words out of Dr. Steidinger's mouth when she walked up to Dr. Burkholder at the airport were: "JoAnn, I'm sorry, but I don't think this dinoflagellate is doing what you say it's doing in its life cycle. We can't seem to get these transformations you say are there, especially the amoeboid stage. I really think you must have a culture contaminant."

Burkholder was disconcerted, but it wasn't long before she gained enlightenment. The next morning, on their way to the lab, Steidinger suddenly remembered that one of her assistants had made a request. After rummaging through her coat pockets, she found a scribbled note: "Ask Dr. Burkholder about the defined recipe for growing this dinoflagellate."

Burkholder frowned. "Haven't you been adding fish?"

"Yes."

"How often?"

"About once a month."

The problem suddenly became clear. Multiple life cycle stages were attained by exposing the cultures *continually* to live fish.

The key now was to get the dino to go through its paces in front of Dr. Steidinger.

Burkholder let Howard Glasgow handle the experiment: nobody else had his touch. His nuances in filtering water, adjusting the amount of aeration in the tank, and adding fish could get the dino to kill fish almost on demand. She would joke that Glasgow could even induce it to perform tricks.

But on this occasion, as though it enjoyed being a spoilsport, the dino refused to come out and play. Nothing Glasgow did could rouse it out of its cystic slumber.

Lamely, Burkholder tried to explain that the organism could be unpredictable. "Laboratory conditions sometimes make it temperamental," she said. But her words sounded hollow even to herself.

A long, uncomfortable silence followed. Burkholder was standing by the sink in her lab, staring at the floor, trying to keep from letting her disappointment get to her. Dr. Steidinger was going to go back, and instead of supporting her, she was going to tell everyone there was nothing to this—

"JoAnn." It was Howard Glasgow. "Something's happening. I think we got it."

She let Dr. Steidinger go over and peer into the microscope. And just moments later, she heard her exclaim, "Oh my God! I can't believe it. Look at this. Look at this!" But Burkholder remained where she was, because she was well aware of what Dr. Steidinger was seeing. After an hour's recalcitrance, the cells were transforming, just as she'd said.

Getting Dr. Steidinger to validate an important component of her research not only brought legitimacy to her other claims about this organism, it also added weight to her overall stature as a scientist. But it wasn't worth much if no one in the scientific community knew about it, so Burkholder came up with a plan. She talked Dr. Steidinger into presenting a joint paper with her at the annual International Conference on Toxic Marine Phytoplankton. She wanted those who had heard her speak in Rhode Island two years earlier to know that her work had progressed, and outrageous as her revelations had seemed to some, now even the renowned Dr. Steidinger was a believer.

Over the following months, Burkholder and Steidinger maintained close contact as they continued their respective efforts to understand this phenomenal creature's uniqueness. Perhaps most startling was the discovery that they had been tricked: they were dealing with not a plant but an *animal*. Under the microscope, the dinoflagellate had looked and acted like a photosynthetic, vegetative cell, but further analysis revealed that although it would sometimes stain the way plants did, it was in fact stealing chloroplasts from other plants, not actually producing them. Indeed, it possessed the complex nutritional needs of an animal.

Deciding on a formal name for this organism also occupied their attention during this period. After a considerable amount of haggling and looking up Latin names for "phantom," "ghost," and even "poltergeist," they settled on *Pfiesteria* (pronounced Feast-*eer*-ee-ah, as in "cafeteria") for its genus name and *piscicida* (pronounced pis-ki-*seed*-ah) for its species name. *Pfiesteria* was a tribute to their friend the late Dr. Lois Pfiester, a pioneer in unraveling the sexual life cycles of freshwater dinoflagellates; and *piscicida* translated as "fish killer."

As the fall conference neared, Dr. Burkholder had every right to feel good about the way things were going. But even as she reminded herself that she was closer than ever to what she'd wanted and should be happy, she found herself tense and restless because something strange was going on with her associate. Howard Glasgow was missing appointments and disavowing conversations, and his work was increasingly sloppy. And when she tried to talk to him about it, he was caustic where he had always been amiable.

Burkholder was traveling a lot, conducting seminars, and when she returned she would ask him, "How's it going?" When he didn't completely ignore the question, he would give her a flippant comeback: "How's what going?"

Obviously something was bothering Glasgow. But when she asked him about it, he acted as if he didn't know what she was talking about. *Has he lost respect for me? Is he no longer interested in the work? Perhaps I've been working him too hard.* She thought all these things. And although he'd never complained about the way she treated him, she decided that must be it. He was feeling unappreciated. And to repay him for all the things he'd done, as well as to give him a boost, she decided to expand the presentation she and Dr. Steidinger were going to give to include him.

The very next day, she urged him to perform some very simple experiments documenting the dino's response to certain nutrients and then to prepare a presentation for the fall conference. Since it was going to be in France this year, she said she would arrange to pay his way, and she encouraged him to make a vacation of it, to take his family along and see Europe.

Glasgow seemed to perk up at the idea, leading her to suspect she'd been correct about what was wrong.

But then it was back to sarcastic comments and not getting along well and spacing out conversations. On numerous occasions she would ask him to do something for her, and when she inquired as to the progress or outcome, he either hadn't got around to acting on it or said he didn't know what she was talking about.

Through July and August she tried her best to stay out of his way, but in September, with the conference just a month away, she felt she had to ask him how his project was coming. What she got was a list of excuses. He'd been busy getting the trailer set up. Feeding the dino had taken a lot of time. He'd been doing this and that, and he just hadn't managed to work up the results.

Because they were having such difficulty communicating, she did not want to add stress by accusing him of procrastination, but she did say she'd like at least to see a graph of his findings. Curtly he told her he'd have something to show her by the end of the week.

On Friday, he brought her a graph that was supposed to document the result of his nutrient experiments, and it was appalling. Glasgow's work had always been sharp and meticulous. What he handed her looked like a sixth-grader's work. If she hadn't known what it was supposed to be, she wouldn't have known what she was looking at.

It's like he's completely fallen apart, she thought.

She knew her disappointment angered him, but she wasn't able to continue to hide her feelings. If he didn't do better than this, it could look really bad for both of them in France.

After he stalked off, she found herself at an utter loss. She didn't know what was going on or what to think. The best she could hope for was that once they got to Europe, the person she had come to count on as a partner and love as a brother would somehow return.

The conference was held in Nantes, a seaport on France's west coast; and the overriding theme that pulled the majority of the presentations together was that human activities were wreaking more havoc on the world's coastal waters than had previously been recognized.

Until this time, while most everyone acknowledged that man-made pollution stressed ecosystems and could create conditions that favored the proliferation of opportunistic organisms like algae, there had been uncertainty about the evidence, which some called incriminating but others insisted was still circumstantial. At this conference, the leaders in the field presented compelling papers that argued for increased eutrophication—the abundant accumulation of nutrients—from domestic, agricultural, and industrial effluent as responsible for a virtual epidemic of coastal algal blooms. Going one step further, they maintained that the repercussions went beyond the devastating financial losses to fishermen and aquaculturists. There were profound human health implications. Studies from the North Sea to India, Spain to Japan, Italy to Australia suggested that waterborne-disease outbreaks were on the rise, poisoning people and in some cases causing deaths.

The information that Dr. Burkholder et al. were presenting about the emergence of a novel pathogen fit perfectly into the bigger picture and once again was a hit at the conference. Burkholder led off, relating how the organism was discovered, the multiple stages of its life cycle, and the linkage that had been made to fish kills. Then Dr. Steidinger spoke, and any doubts about the legitimacy of Dr. Burkholder's statements were dispelled when one of the world's foremost authorities on toxic dinoflagellates stated for the record: "I've seen these transformations with my own eyes, and it's the most amazing thing I've come across in my thirty years as a scientist."

As JoAnn Burkholder sat in the audience, listening to those words, she thought, *What I've given to hear you say that!* But she was unable truly to enjoy the moment, because her stomach was in knots with worry about Howard Glasgow, who had become steadily worse. He was sullen and morose much of the time and no longer seemed to care what she thought. Two weeks before the conference was to start, he still had not prepared his talk, so she'd had to redraw his graphs and put his slides together. Up until a half hour before their segment was scheduled, she'd been in his hotel room, coaching him. "Here's what you show," she'd said, putting a slide or a graph in front of him, "and here's what you say."

It had seemed to her that he was trying to concentrate but just wasn't able to retain the information, and he had resorted to writing down every word she said. As they rehearsed, she'd found herself thinking, *This has to be the worst case of stage fright I've ever seen. Maybe he's not ready for this and I should have started him out with a smaller conference.*

When Dr. Steidinger finished, it was Glasgow's turn, and he walked very slowly to the stage. After rustling some papers, he started to speak. He read straight from his notes, speaking slowly, his presentation broken by frequent pauses, as though he kept losing his place. Every second of silence seemed ten times as long to Burkholder, and each time he stopped, she held her breath.

Then, about two-thirds of the way through his talk, he just couldn't seem to hold his thoughts together any longer. The silence was deafening. Although she knew it would be devastating to him, Dr. Burkholder was just about ready to get up and join him onstage, when he started up again; she sank back in her chair.

It was painful to witness, but in the end he pulled it off. When he stepped down from the stage and took a seat beside her, she grabbed his hand and held it tight. "You were just great," she whispered, and she could tell he was as relieved as she was.

After a break, there was a half hour set aside for questions for everyone who had spoken that day. If further indication was needed of how interesting Dr. Burkholder's contingent had been, it was provided when a single question was asked of all the other presenters, then so many were asked about *Pfiesteria* that the moderator finally asked Dr. Burkholder to remain onstage and field questions for the rest of the period allotted.

The first few questions were fairly basic—How did the fish look when they died?—and she was able to answer them straightforwardly. But she had a little more trouble with the questions posed by a man in the audience with a thick Scottish brogue whom she recognized as Dr. Jeff Wright, a Canadian scientist who had led a team of colleagues in solving the mystery of a series of illnesses and deaths on Prince Edward Island in the late 1980s, tying them to shellfish poisoning by an organism that had always been considered benign and beneficial.

"You've told us about an organism that has killed literally millions of fish. Do you have any evidence that it is causing problems for human health?"

It was Burkholder's turn to pause, as she considered what was prudent for her to say.

"To be honest, we don't know much about what effects it has on human health. We think we've seen some problems in a laboratory setting, but at this point they're more anecdotal than clinical. Beyond that, we can't say much."

She hoped that would satisfy his curiosity, but the man remained standing.

"Would you mind describing what kind of effects those were?"

"We think it has caused short-term memory loss. And there have been other symptoms reported, such as disorientation, abdominal pains, and eye irritation. I can't say much more than that, but that is what happened to one individual working with it."

She glanced around, hoping to break the line of questioning by calling on someone else.

"Was that somebody from your lab?"

She smiled. "Yes. It was me."

When the buzz in the audience subsided, she added, "But fortunately the symptoms subsided relatively quickly. As for the short-term memory loss, it appears to be reversible."

That wasn't entirely true. In fact, she continued to suffer from lingering health problems that she associated with her exposure to the dino. Whenever she tried to do strenuous exercise, she came down with bronchitis. Twice she had been diagnosed with pneumonia, which she'd never had before. Regarding her memory, as she later said, "I used to be razor sharp. I had a photographic memory. I had an edge. A little bit, but just enough to put me out in front. And that little edge, when you're already at the top, translates into miles. I could sit in a room, and as my major professor described it when he compared me to other students, I ran rings around other students, and I didn't know I was doing it until I realized nobody else remembered what I was regurgitating. . . . But I'd lost it. Maybe nobody else would notice it, but I did. I found myself having to compensate. Work harder. Write more things down. Which I detested."

The rest of the conference was almost anticlimactic, but not without its memorable moments. After Howard Glasgow departed on a sightseeing trip with his wife and six-year-old daughter, allowing Burkholder to think about something else, she found herself the focus of the kind of attention that disturbed her most about the toxic-dinoflagellate crowd.

Since she'd flown to Miami two years previously to consult with Dan Baden, just the thought of the toxicologist and how he all but laughed her off quickened her pulse. She had noticed him sitting in the audience during her talk and given him a flat look, and he had lowered his eyes. But later he caught up with her, and he couldn't have been more respectful, even contrite, as he said, "Look, I know we got off on the wrong foot, but I want to make up for it. I believe what you're saying now. Let's let bygones be bygones. How about giving me another chance?"

Caught off guard by the complicated irony of his reversal—it was important for the toxin to be characterized, and he was one of the best toxicologists in the world, but she was proud and wanted to be treated as a colleague—she put him off, telling him she would think about it.

A banquet was held the final night, and she attended alone, milling among the scientists and their spouses, hoping to meet people who were doing interesting work or could talk about what was happening elsewhere. After wandering aimlessly with a glass of wine in hand, she began to drift in the direction of the dining room. She was looking around for Jeff Wright, the inquisitive Canadian scientist—she wanted to sit at the same table with him so she could ask him about how he'd cracked the shellfish poisoning case—when a suavely handsome fellow, dressed in casual sports clothes, walked up and told her how much he had enjoyed her presentation. She thanked him and, spotting Jeff Wright, began to edge toward him. But her admirer followed, continuing the kind of small talk that made her feel he was trying to charm her. The way it worked out, they all ended up sitting at the same table, Jeff Wright across from her, the persistent stranger beside her. And it wasn't long before she discovered why he was so interested in her.

He started out by asking, "Who's doing the toxin analysis?"

When she said she was considering giving Dan Baden one more chance with it, he said, "Why don't you let me do it?"

At that point she turned and looked directly at him. "And who are you?"

His name didn't register as strongly as his place of employment. He was a toxicologist with USAMRID, the United States Army Medical Research Institute of Infectious Diseases, at Fort Detrick, Maryland.

She didn't know much about what went on at Fort Detrick, but what she'd heard made her uneasy. The word was that the army worked with a lot of deadly toxins there and were very secretive about what they did with them. She'd been told they had found a way to use the toxins associated with red tides for military purposes, and that when civilian scientists interested in developing deterrents to red tides had requested information from the army, they'd been told the information was classified.

The conversation went on a while longer before the fellow came up with what he thought was a perfect solution. "Why don't you let Baden work on the toxin in the water and let me handle the aerosolization?"

"Why are you interested in that part?" she asked.

The fellow from Fort Detrick grinned. "If what you say about short-term memory loss is true, can you imagine its value to biological warfare? We could hit people with that, and they wouldn't remember what happened."

Later she would think maybe he'd been joking. But at the time, she remembers, she concentrated on the glass in front of her and wished it were full of Jack Daniel's instead of white wine.

At precisely that moment, Jeff Wright leaned forward and asked, "JoAnn, would you like to dance?"

It was a fast waltz, music she'd never dance to normally. But under the circumstances, it was the perfect rhythm for a perfectly timed rescue.

～～ ～ ～

When Howard Glasgow returned from Europe, he seemed to be much improved. A vacation appeared to have been what he needed to raise his spirits. He came into the lab joking and laughing, and people around him were saying the old Howard was back.

Burkholder had gone to the coast on a field trip with one of her classes, and she had left a note behind for him, saying she had decided to give Dan Baden one more chance, so they needed to accumulate as much toxin as they could as fast as they could. If the culture had been maintained correctly while they were gone, that would have been no problem. But Glasgow found that the undergraduate who was supposed to have been feeding the dino had been afraid to go into the "hot" (toxiculture) room, which meant they had to grow it again from scratch. Glasgow was in the lab for thirty hours that weekend—fifteen hours a day—feeding fish to the dino and trying to get it to produce toxin. By Tuesday he was surly and temperamental, and people were saying, *Boy, that was short-lived.*

Burkholder had begun to suspect that there might be a connection between Glasgow's behavior and the dinoflagellate. In his forgetfulness and disorientation, in the way he would fly into a rage over nothing, she recognized some of the symptoms she had experienced, though on a reduced scale. But it didn't make sense. She didn't understand how it could happen. There were only two ten-gallon tanks in the hot room where toxin was being grown, and the exhaust fans were venting the air twelve times an hour. Besides, Glasgow had always been so careful and safety conscious.

It was one of the most helpless feelings she'd ever experienced: watching, mystified, as someone she cared about deteriorated in front of her eyes.

There were eight or nine students working in the lab during this period, and they were getting apprehensive. When some of them reported to Dr. Burkholder that Glasgow's behavior had become a source of concern to them as well, she found herself thinking the unthinkable. Unless things changed, and soon, she was going to have to let Glasgow go.

She wanted to postpone that decision, particularly with Christmas just a few weeks away. But the next time Glasgow came to her office, she did say she wanted him to cut back on his hours of working with the toxic dinoflagellate. To spend no more than three hours at a time in the hot room, and never alone. "I do not want you in the facility as much as you've been there," she told him. "I want you to take some time off and make sure you're all right."

When he started to protest, she said, "I'll cover for you."

Howard Glasgow had persistently denied that there was anything wrong with him. During this conversation, however, he finally admitted that he had experienced some confusion lately. He also said he'd been afraid to admit it to her earlier, and then he'd forgotten about the whole matter.

"Howard, go home. Please."

With resignation, he said he would, and he rose to leave. Noting the shuffle in his walk as he left her office, she thought it was like watching someone with Parkinson's disease, and it broke her heart to see him that way.

This conversation took place on December 6. Burkholder remembers the date clearly because of what happened next. A group of students from her lab came bounding in from a long, cold day taking water samples on the Neuse River. "Hey. How about a pizza?" she said, and a cheer went up.

"I have to stop by the trailer for a moment, and then I'll pick up pizza for everybody."

Raleigh was lit up like a scoreboard with Christmas lights. As she drove, she thought about the shopping she had to do. Glasgow had said he'd changed the fish before coming to her office, but she wanted to check for herself. Given his condition, she had a hard time believing he could have done it.

When she arrived at the trailer, she followed the same safety procedures she expected from her lab assistants; and after donning a lab coat, gloves, and a respirator, she opened the door to the hot room and flicked on the lights.

The sight that greeted her was as shocking as a shriek in the dark. It was a chamber of horrors. Open-topped aquariums were bubbling. Dead fish were scattered across the floor, some caked dry, indicating they'd been there at least a day. This, an hour after Howard Glasgow had said he'd been at the facility.

It took her almost three hours to clean up, and she knew she was going to have to come back and bleach the floors and walls and shut down the aquariums because of the probability of cross-contamination.

She next saw Glasgow two days later. When she told him what she had found at the facility and that he was not to be allowed back until they figured out what was wrong with him, he refused to believe her or to accept her judgment.

"You can't keep me out," he said, getting emotional. "I'm going to continue working on this. It's not for you to tell me I can't. It's not fair of you to do this to me after all I've done."

He was sitting on the opposite side of her desk, and she looked at him. He had a dull sheen to his eyes, and his mouth hung partly open. Compassion swept over her, and yet she could not let him go back, for his own sake. Whatever was happening, it clearly was endangering his health. *What do I do?* she asked herself. *How do I handle this?*

An idea occurred to her. A test to see if this was as bad as she thought it was.

"Howard, I understand what you're saying, and I respect you for that. I realize you want to continue with your work, but you have to respect my opinion too. You cannot ask me to watch you go over to that facility and get sick and possibly die. I'm not going to do it. It's not fair of you to ask me to do it."

He sat there staring straight ahead, saying nothing. Tears suddenly bloomed in her eyes. When she recovered her composure, she continued.

"Howard you have two daughters to think about. You have a family you care about. You can't ask them to allow you to do this either."

Still he sat there, saying nothing, registering no reaction.

"This reminds me: Did you ever read the Nick Adams stories by Ernest Hemingway? Do you know any of those stories?"

His expression sagged, and she wondered if he was listening.

"There's this one story where Nick Adams attends his father's funeral, and his father happened to be a scientist. And one of his father's colleagues comes up to Nick and says, 'What a terrible loss. What a blow to science.' And you know what the son said? He said, 'Oh, science took it awfully well.' "

She paused, giving him a moment to think.

"Do you understand the significance of that?"

Nothing. Then, "I guess I can understand what you're saying. But I really have to keep doing this. I'll admit I feel like hell, but . . ." His words, along with the thought, faded away.

"Howard, you haven't been yourself for a while now." Burkholder went on like that for a couple of minutes before she came to the point. "Remember the Nick Adams story?"

He looked at her. And very slowly, very sadly, he shook his head.

Burkholder lifted her eyes to a far wall and took a deep breath, trying not to cry. Then she got up and walked around her desk and hugged Howard Glasgow.

"I want you to get the hell out of here and go home," she said. "I don't care if you hate me. I'm padlocking the trailer, and you're not going back until a doctor can figure out what is wrong. I want you to get *well.*"

The day Howard Glasgow was offered an assistant's position at Dr. Burkholder's laboratory, his wife, Aileen, a green-eyed, chestnut-haired beauty from southern Georgia, had called all their friends and invited them to a backyard barbecue to celebrate. She knew how important this was to him. He had talked about his ambition to become a scientist someday almost from the time they'd met as college students at the University of North Carolina at Wilmington. In the twenty years since, she'd seen one thing after another prevent him from pursuing his dream: the unexpected arrival of a daughter, Amanda, while they were still students; the loss of a scholarship to medical school due to federal cutbacks by the Reagan administration. Another surprise, whom they named Lesley, had been followed by the financial fiasco with his father.

Her husband had not had a plan when he said to his father, "Enough." He didn't even have a ride home. When he handed the ring of office keys to his father, included were those to the leased service van that had doubled as a personal vehicle.

There were those who thought it had been hasty of Howard, with a wife and two daughters, to leave one job before he'd lined up another, but Aileen Glasgow was nothing if not supportive. She knew her husband could have gone to work the next day if he'd been willing to remain in the electronics field, but she also knew that he saw this as a delayed opportunity to pursue his first love.

By nature, Aileen was a cheerfully optimistic person. In the medical center where she worked as an office nurse, they sometimes called her Tinker Bell, because nothing seemed to get her down. And over six months that she watched Howard pore over the newspaper each morning, responding to every science-related job advertisement, and mail applications to every North Carolina company that had a scientific research department, and not get a single call back, her spirits never flagged. Even though their savings were running low, she maintained a religious faith that things would work out.

The interview at North Carolina State with Dr. Burkholder had been his first in-person interview. And even though it was only part-time employment and the pay was not nearly enough to live on, that didn't matter to her, because she was confident things would change as soon as Dr. Burkholder saw how smart and organized her husband was, what a hard worker. Howard was the brightest person Aileen knew. He could remember specific problems on tests he'd taken in college when she didn't even remember taking the course. And he'd always been obsessive about work. He was eleven when his father left the family, and in order to bring money into the household, Howard had gone to work at a Red & White supermarket, stocking shelves from five to eight every morning, walking to school, then working again in the afternoon and on weekends.

Aileen had been proud of her husband's accomplishments as a businessman, but she threw a party for him because she knew this was the start of his brilliant career as a scientist.

Workaholic though he was, Howard had not been one to bring his work into the house, so Aileen had not known a great deal about what he was doing in Dr. Burkholder's lab with the dinoflagellate. She did know he was excited about the project, because he said if they could identify what was killing fish in the estuaries, it was something that could make an immediate difference, whereas most rewards in science were slow in coming. When he told her it was toxic, that didn't worry her, because she knew her husband was a careful man. As for his "episodes," he talked about them as if someone had hit his funny bone. It hurt at the time, but you could laugh about it later. "Be careful," she had reminded him. "Oh, I will," he assured her, and she didn't think any more about it.

In retrospect, it would be hard for Aileen to put her finger on the first time Howard acted in a way she would later blame on the dino. She would remember that he was irritable during the early months of 1993, but she knew he was going through a series of frustrations in getting the new laboratory set up; and he was working long hours, so he wasn't getting much sleep. She tended to think of his mood swings in terms of his dedication to his work, because she knew that "Howard does not think of what he's doing as an eight-to-five job. If he was doing an experiment and it wasn't coming out, he would do it over and over and not give up until he found what he was looking for."

If she had to cite a beginning, she guessed it was their European trip.

From the start, he was different. When they were in college together, he would sometimes get nervous and tense before he had to give a talk, and he had always wanted her to be there for moral support. But not in Nantes. He didn't want her anywhere near the conference. The day of his talk, he told her to take Lesley—they brought their six-year-old with them—and visit the city's famous cathedral. Which she did, spending most of her time on her knees, lighting candles and praying for his speech to go well. That evening over dinner, when she asked him how it had gone and what questions he'd been asked, she was surprised that he couldn't give her an answer. He said he didn't remember.

That was odd, but she figured maybe he was just so relieved it was over he'd blocked it out. After that he seemed to relax, and everything was wonderful as they drove around France in a rented car, taking pictures and stopping at village inns.

Because the city of Glasgow was the origin of his surname, Howard had wanted to see Scotland. From London they took the train north, and for the first leg of the trip everything went fine. Then Aileen turned to Howard and asked, "Would you like to write some postcards now?"

It was a perfectly reasonable request. They still had three or four hours before they arrived in Edinburgh, and they weren't doing anything else. And if that wasn't what he wanted to do, he could just have said, *No, I think I'll take a nap.* Or, *I don't feel like doing that right now.* There was no reason for him to scream, "I'm not doing any damn postcards."

She was shocked and embarrassed. People were staring at them. After sitting in silence for a while, she moved to another seat to read a story to her daughter. But the incident remained with her because it was so uncalled for, and because when she asked Howard later to please explain his rudeness, he acted as if he didn't know what she was talking about.

The red lesions started to appear several weeks after they returned from France. He came home one day with a red rash on his arm and said the Plexiglas top of an aquarium had slipped and water saturated with the dino had splashed over his glove. By evening it had turned into a running sore. Aileen treated the wound with an antibacterial ointment, but over the next several days it spread: to his back, his chest, his feet.

His disorientation began four or five days after that. He would pour himself a glass of orange juice, and instead of returning the container to the refrigerator, he would put it in the dish cabinet. A little while later, he said he was going to make cocoa, and he turned the burner on and put a newspaper on the stove instead of a pot of water. Fortunately Aileen caught his mistake, just as the paper started to burn.

"What are you doing?" she cried. "Have you lost your mind?"

At the time, there always seemed to be an explanation handy. He put the juice in the cupboard because he wasn't paying attention. He wasn't the only one who had trouble with the stove—she was always turning on the wrong burner. And she even rationalized his doing something completely nonsensical—such as the snowy morning when he said he had to get to work early and was halfway to the car when she called after him, "Howard, are you forgetting something?" He turned and peered quizzically at her. She looked down at his feet and so did he, appearing to notice for the first time that he was standing there in his socks. *Oh, that's just Howard being a kidder,* she thought. He was constantly teasing the girls: hiding behind doors and jumping out and funny stuff like that. So maybe he was doing this just to see if she would notice.

But then came the rages. Frightening, inexplicable blowups. Like the night he woke up yelling at her not to pile the vacuum cleaner cord on top but to coil it. She didn't know which of them was having a bad dream, but when she questioned him about it the next morning, he claimed that she must have had a nightmare. Another time, she was at the piano, in the middle of a song, when he barked at her to stop playing it wrong. "You're using too much pedal," he complained. She was baffled. She'd played it the same way for twenty years.

Then noises began to grate on him. Suddenly he couldn't stand to be in the house when the dishwasher was running. Aileen agreed, the sound of a dishwasher was enough to irritate anybody, but after listening to it for eight years, he should have been used to it.

And friends would call, breathlessly, to say they'd just seen Howard on the highway, eyes straight ahead, both hands on the steering wheel, driving like a maniac.

When Aileen began to implicate the dinoflagellate he was working with, he denied it was even possible. "We're being more careful now. We've got this new lab, and we've upped the safety stuff."

But then he was also denying the abrupt temper losses. And he would deny he was forgetting things. This, when you couldn't tell him anything and expect him to remember it. They were supposed to go to a PTA meeting one Friday night, and Howard was late coming home from work. When she called the trailer to see if he might still be there, he was. "You didn't tell me anything about a meeting," he said. She couldn't believe what she was hearing. They'd been planning it for a week.

By now he was also having severe physical symptoms: hammering headaches, high blood pressure, and an irregular heart rate. She wanted him to go see a doctor, but she was walking on eggs around him as it was and didn't want to trigger an outburst. She was trying to think how to trick him into going, when he caught himself getting mad at someone in the lab and finally seemed to realize that it wasn't just a matter of people blowing things out of proportion; he had overreacted and been unable to control himself. That evening he said to Aileen, "I think maybe something is wrong. I think I should go to a doctor." She made an appointment with an internist.

Aileen had to work that day, so she let him go by himself. Afterward, Howard came by the medical center, and she could tell he was upset. "He wants me to have an MRI of the brain," he said sadly.

"An MRI? Why on earth?"

Howard had been given a series of tests and had flunked every one of them. The doctor had written ten words on a sheet of his stationery and shown it to him and asked him how many words he could remember, and all he could recall was the letterhead. Then the doctor had held his finger up and asked him to follow it with his own, and he couldn't do it.

The day of the MRI, Howard was supposed to come home from work early so they could go to the hospital together. Aileen had told him twenty-five times not to be late, because she knew these things were scheduled tightly. She waited and waited, and he didn't come and didn't come, and at last she called work and they said he'd left hours before.

Eventually he showed up, and he was ashen.

"What's the matter? What happened?" she asked.

He answered in a small voice. "I got lost. I couldn't remember how to get home."

He said he'd stopped once to ask for directions, but because he couldn't remember his address or even his telephone number, no one could help him. He described his confusion as "like a fog" and said he just kept driving until the fog lifted.

Aileen wanted to drive him to the hospital, but he wouldn't let her. Since the day they'd met, *he* drove them. So she let him drive, but she kept a close eye on him.

He was okay until they got to the highway. Then he turned into a little strip mall and asked her, "Which is the hospital entrance?"

"Howard, this is the mall. See, it says North Duke Mall."

He looked, and she could tell he was bewildered. And it wasn't as if they were wandering through unfamiliar territory; the hospital wasn't more than a mile from their house.

"Go back out," she told him. "It's just a little ways down the road."

When they finally got to the emergency room, where after-hours patients were admitted, the woman who checked them in started asking Howard questions. Aileen had always hated it when medical receptionists talked in front of patients as if they weren't there, so she wasn't going to answer for her husband—until the woman asked Howard his name and he didn't answer. When she asked him again, he just looked at her as if he didn't have the faintest idea what she was talking about. So from then on Aileen answered the questions.

After the check-in questionnaire was completed, the receptionist gave them involved directions to the MRI facility.

Glancing at Howard, Aileen could see him seize up with fear. "That's okay. We'll find it," she said with a smile.

Aileen led the way, and Howard followed. A female technician came into the hall to greet them. "Hi, how are you?" she said in a friendly, chatty way.

"I'm having trouble with my memory," Howard told her, and the woman laughed. "Oh, aren't we all?" Giving Aileen a wink, she said, "Now you just come with me, and everything will be all right."

After leading Howard by the arm into another room, the technician was gone just a few minutes before she returned to apologize.

"I am so sorry, Mrs. Glasgow. He really is having problems, isn't he? I didn't want you to think I was making light of his condition."

"Don't worry," Aileen said. "I know he looks perfectly normal until you start talking to him. But he will cooperate with you. You just have to keep telling him over and over, or else he won't remember."

After the MRI, Aileen drove home, and as if he'd forgotten that he'd ever done otherwise, Howard sat contentedly in the passenger seat. Holding up copies of the film, he was able to joke, "Look. In case anybody might think otherwise, here's proof I have a brain."

Aileen's mind raced on a solitary track. She was worried sick, and the last thing she was thinking about was dinoflagellates, because there were so many possibilities that made more sense than some exotic organism no one knew much about. She was thinking brain tumor and neurological disease and multiple sclerosis, all of which were very likely explanations for his symptoms.

The next day, he was worse. He stuttered. He would begin sentences and not finish them. Then he stopped speaking altogether, and she was especially frightened, because always before she could ask him, "Does your head hurt? Is anything wrong?" and receive an answer.

After a long weekend during which his mental deterioration was so alarming and so rapid that Aileen considered taking him to the emergency room, they met with a highly respected neurologist from Duke University Medical Center, Donald Schmechel. While he refrained from making an on-the-spot diagnosis, on the basis of a preliminary examination Dr. Schmechel said that Howard's symptoms suggested serious neurological injury. He told Aileen that her husband should not drive and should not go back to work. He said he wanted to see Howard again for a battery of neuropsychological tests, because there was a whole list of things they had to rule out before they would know what the source of the problem was.

If there was anything positive about their situation, it was Aileen's being uniquely prepared for the role she would be playing. Her father had been a surgeon, so she had grown up around medical emergencies. She could remember sitting in his office at the age of three and watching him sew people up. More recently, when his nurse was away, she had done lab work for him. As well, she was no stranger to mental disorders. After an accidental fall fractured her skull, Aileen's mother had suffered from a permanent brain malfunction that caused her to confuse words. Like Howard, she had complained of a fog that came and went. As if that wasn't enough, Aileen's grandmother, who lived with them for years while Aileen was growing up, experienced hallucinations as a result of a series of strokes. Fortunately they were happy hallucinations—she didn't see people sneaking into her room to kill her; she saw horses on roofs. As a child, Aileen would run to the house next door and say, "Miss Hattie, my grandmother thinks there's a horse on your roof. Can you please come and talk to her." And Hattie Bragg, a sweet, elderly soul, would oblige her. "I am so sorry," she would say to Aileen's grandmother. "I had forgotten that horse was on my roof. I'll go get him down right now."

So in a way, when these latest strange circumstances entered her household, Aileen was better equipped than most wives to handle them. Told that Howard could not go to work, however, she almost panicked. *Oh my God, what am I going to do?* she thought, because all her husband seemed to care about was work. He didn't really have any at-home interests or hobbies. And as sympathetic as she was toward his plight, over the previous month she had found her patience sorely tested.

But what happened was strange. Once he was told he couldn't go back to work and was not allowed to drive anywhere on his own, something inside Howard appeared to relax, and he became a completely docile presence around the house. It was even kind of nice, Aileen would remember, "because even though he wasn't exactly with us, he was home and it was Christmas."

To keep him busy, Aileen tried to put him to work around the house. When it was pleasant out, she would suggest that he putter around the yard, raking leaves; and when it was cold, she would ask him if he felt like doing indoor projects. That was how she got him to wallpaper the bathroom—something he'd always said he was going to do but never got around to.

Hopeful that this was a sign he was getting better, she suggested a larger household project, painting the upstairs bedroom. No problem, Howard said. His brother Mark came over to give him a hand, and they had just about finished rolling the first coat on the walls when little Lesley came into the room to show off her new blue jeans. They were her very first pair, and striking a model's pose, with one hip thrown out to the side and her hands stuck in her back pockets, she said, "Look, Daddy."

And Howard exploded, screaming at her to get her hands out of her pockets.

Nothing he got mad about during this period made sense, so Aileen recognized it for what it was—another pointless tantrum. But it was the first time his brother had seen him like this, and it scared him silly. "I'll take him home with me," Mark whispered to Aileen. "He can spend the night with me. It's not safe for you to be here alone with him."

Aileen reassured him. "In five minutes he won't even remember what he said. He'll be perfectly okay."

And he was, leaving Aileen to wonder whether it was just a coincidence or if there was something in the paint fumes that triggered the episode.

It was toward the end of the holidays that Aileen's worries began to expand to concern about their economic well-being. After completing an extensive series of tests, Dr. Schmechel had determined that indeed there was brain damage. But beyond that, the most he had been able to say to her was: "I feel the dinoflagellate is the cause of Howard's problems, but it's impossible to say for sure. The trouble is that even though we've ruled out just about everything else, nobody we know of has experienced anything remotely similar."

Insurance would probably cover most of Howard's medical costs. It was the loss of income that worried Aileen; and they might have trouble collecting workmen's comp if they couldn't prove her husband's debilitating illness was work related.

Aileen was still working as a nurse, but she had lost a great deal of pay during this period. Even though Amanda, their oldest, would sometimes keep an eye on her dad, or an elderly retired couple who lived across the street would come and sit with him, Aileen had had to take a lot of time off from work; Howard could not be left alone. But as bleak as their financial situation looked, the true tragedy was how, in the pursuit of science, Howard Glasgow may have damaged permanently his chances of ever becoming a scientist.

~~ ~~ ~~

After Dr. Burkholder admitted during the extended Q/A period at the end of her talk in France that she might have been affected physically by working around the toxin emitted by *Pfiesteria,* colleagues had come up to her and commended her on her courage. More people in their field should be willing to come forward and acknowledge potential risks, they said. At the time, she wasn't sure what they meant; she'd merely answered honestly, not bravely. But comprehension wasn't long in coming.

Shortly after her return from France, she began to receive terse calls, letters, and faxes from an official at the National Marine Fisheries Service in Charleston, South Carolina. This was the group that reviewed and awarded grants for the National Oceanic and Atmospheric Administration, from which she had received almost $25,000 in September to pursue research into more environmental characteristics of *Pfiesteria.* Because of her comments about health effects, even though they were anecdotal, concern was being expressed about the working conditions in her lab. She was asked to describe in detail the safety precautions she was taking to protect herself and her assistants from the toxin. She was asked for a copy of the trailer plans for review. She was asked to write up a chronology of everything she could remember that related to possible human health effects from this organism in her lab.

At first Burkholder was receptive to these inquiries, as well as any advice the NMFS could give her. "I am always interested in making a good situation better," she wrote back. And she maintained that stance until she received a fax notifying her that her funding had been suspended and instructing her to cease all research with the dinoflagellate until a team of experts could come to Raleigh, inspect her facilities, and review her safety protocols.

In fact, she had already beaten them to it. Fearing Howard Glasgow's illness could somehow have been related to the dino, she had padlocked the trailer, allowing no one inside but herself—and only to properly disinfect the trailer interior and shut down all the aquariums. Until she determined what had happened to Glasgow and how it had happened, she did not intend to reopen the lab.

But taking that voluntary step was very different from having her funding frozen and receiving orders to suspend operations pending a site visit. She had never heard of such a thing happening, and it struck her as punitive. When she asked for justification, she was told that oral reports had come in that suggested her facility was dangerous and her safety precautions were inadequate, and as the agency funding her, they were concerned about liability.

She knew that was highly questionable. If liability were an issue— and she was absolutely certain it was not—it would be the university that would be held responsible, not the granting agency. So there had to be some other explanation.

She was still trying to put it together when she received a call from an officer with a federal agency in Washington, D.C., that had provided funds for her research at an earlier date.

"What's going on down there?" he asked, as if he expected her to know what he was talking about.

"What do you mean?" she asked.

"Do you have any idea what is being said about you by the people at NMFS Charleston?"

She tensed. "No. What's being said?"

"Well, I just came from a meeting at which they said they were investigating you because you can't be trusted to work safely with toxic dinoflagellates."

"That's bullshit!"

"That's not all they said. They're also recommending that no one else take a chance with you. In essence, there seems to be an effort going on here to blacklist your funding. What the hell's up?"

Now she understood. An administrator at the National Marine Fisheries Service in Charleston who had promised her at a seminar she'd given two years earlier that he would find a way to obtain her *Pfiesteria* culture and take over the toxic research—"Just watch," he had said—had at last found a situation he could manipulate in such a way as to keep his promise. If the site visit went the way he wanted, she had no doubt but that the team of investigators would come to town, condemn her facilities, brand her as incompetent, and recommend that her work be transferred to NMFS Charleston.

The six handpicked experts arrived in Raleigh in early January of 1994. One was from the Centers for Disease Control, one from Fort Detrick, two from Florida, and two from NMFS Charleston. They made their way to the Brownestone Hotel, which was within walking distance of the North Carolina State campus, and the next morning they were picked up in a university van. JoAnn Burkholder was driving, and everything about her conveyed her mood. She was dressed in black, except for a blue blazer. She had a dark look in her eye and a cold cast to her face, and she did not get out and greet them but let them load themselves. She was bitterly resentful about what she felt was a charade, and wanted them to know it.

At the trailer, she guided them through the facility, her voice matter-of-fact. The team looked around and jotted notes, and forty-five minutes later she drove them back to the hotel, where they had lunch and discussed the situation privately.

That afternoon a formal meeting was held around a long table in a hall on campus, where she had to give yet another synopsis of her precautions. She made no effort to disguise her contempt for these proceedings. Her statement was followed by that of personnel from the N.C. State Environmental Health and Safety Center and university officials, all of whom testified that she had come to them for advice on numerous occasions and had done all that could reasonably be expected of an investigator trying her best to figure out appropriate safety precautions when working around an organism that no one knew anything about.

By the end of the site visit, it was obvious to JoAnn that the affair had more to do with politics than precautions. Several recommendations for improvements to her culture facility were made, but none were beyond what she had already considered. They certainly didn't justify the expense of bringing in six scientists from around the country. Nor were they the kind that would threaten her right to continue research on this organism. In fact, the site visit team, while stipulating that certain additional safety measures would have to be in place before the funding freeze would be lifted, concluded that the health of her lab workers had for the most part been responsibly protected by the safeguards she had put in place.

And so, what she suspected was an attempt at a research coup failed, but not without taking a toll. Burkholder was left feeling drained, violated, and exposed. Her integrity as a scientist had been attacked. She'd been made to dance for people she did not respect. She would have no way of knowing, until she applied for more grants, if there would be any fallout from the failed attempt to blacklist her. And she felt she'd been put through all this because of scientific greed.

<hr />

While Dr. Burkholder emerged from the site visit more or less intact, throughout it she continued to wonder what had happened with Howard Glasgow. If they were taking all these precautions, and if six experts couldn't more seriously fault her protocols, then she could not understand how the dino could have been involved. But what else could it be? She could not stop thinking about it. Miserable without an answer, late at night she began to drink heavily—Manhattans, straight up. But she still couldn't sleep. Glasgow's condition, and the prospect that it might be related to *Pfiesteria*, haunted her.

Then one evening about a week after the site visit, she found herself sitting alone on her apartment balcony, wearing a leather jacket and fur-lined boots, staring through a frozen fog at the thoroughfare below and the big apartment buildings across the way, thinking and drinking,

drinking and thinking. Finally, around two in the morning, she rose and stumbled into the bedroom, but before she fell asleep she put her subconscious to work. It was a technique she'd used before to come up with solutions to problems: just before dozing off, she would read something or concentrate on a topic, and the next morning she would wake up with an idea. On this occasion, she thought the whole thing over one more time before drifting into a restless sleep.

Usually she didn't remember her dreams, and she wasn't sure whether this was a dream or a memory, but something triggered the recollection of a comment Howard had once made to her in passing: "You know, sometimes I think I can smell that stuff even when I'm in the cold room. But I know it can't be true." No more than two hours after closing her eyes, Burkholder sat bolt upright in bed, remembering his words and thinking, *That's it. There's got to be something wrong with the ventilation system. There's no other way.*

She turned and looked at the clock on the bedside table. It was four in the morning, too early to call anybody, but she knew she would never get back to sleep, so with three or four hours on her hands, she got up, dressed, and got in her car. She drove to a little stream she had once taken samples from, about fifteen miles from her house, and she sat in the dark listening to Led Zeppelin and Melissa Etheridge tapes with the volume turned as loud as she could stand it, thinking, *I've got it—I have the answer,* until the sun came up and she knew the campus Health and Safety Center would be open for business.

An environmental safety officer by the name of Bruce Macdonald fielded her call. When she told him she wanted the airflow in the culture facility checked and asked how soon he could do it, he explained to her that he wouldn't be able to do a thorough examination for a couple of days, but he could do a smoke test later that morning, which would be a start.

"At least that will tell us if the direction is right, won't it?" she asked.

"Oh, yeah," he replied.

"Okay. Let's do it."

She had called from her office, and when she arrived at the trailer five minutes later, Macdonald pulled up beside her. She let him inside, and they went directly into the hot room. Well before the site visit, she had bleached everything down and removed the aquariums, so they didn't need to take any precautions, and now she closed the door between the hot and cold rooms and switched on the exhaust fans to duplicate the conditions under which Glasgow had been working.

Going briskly about his business, Macdonald said, "We can test this in several ways," as he pulled out a smoke flare—a plastic tube that contained titanium and chloride, which produced smoke when you squeezed one end—and set it on the floor just inside the hot room. Then he walked over to one of the ceiling fans, which was humming as loud as a turbine, took out a Kleenex, reached his arm up, and held it there a moment before letting go.

If the fans were working properly, there should have been enough suction to keep the tissue in place, but it parachuted to the floor.

"Hmm. That's not so good," Macdonald murmured, sticking the Kleenex up to another ceiling fan and watching it do the same.

As Burkholder took this in, she shook her head in disgust, thinking, *So much for twelve air exchanges an hour.* Then her eyes moved from the ceiling to the floor. The value of a smoke test was that it allowed you to see the direction in which air was moving, and it was doing this clearly. The problem was that the smoke was being drawn *under the closed door.*

"Oh, Christ," she said softly as she stared down at the floor. "Look at this."

Macdonald came and stood beside her. After a moment, he opened the door, and together they watched as the smoke billowed out of the hot room, turning the cold room cloudy.

Now she knew what had happened to Howard Glasgow. The exhaust fans weren't doing the job of ensuring air exchanges, the door separating the hot and cold rooms had not been properly cut, and as a result, the airflow in the facility was reversed. Contaminated air had been flowing into the area that was supposed to be sealed off. Glasgow had thought he was working in a safe area, when in reality he was being exposed to high levels of toxin without any protection whatsoever.

More flaws in the system would be discovered. They would find that the duct system from the hot room, instead of venting outside, actually recirculated air directly into Howard Glasgow's private office. But right now Burkholder knew all she needed to know. The trailer had been improperly constructed, and Howard Glasgow had been slowly and systematically poisoned over the previous five months by the dino-flagellate's neurotoxic aerosol.

"Do you want me to do anything else?" Macdonald asked her.

Her voice was icy. "No. I have the answer I needed."

PART THREE

"Watch out for Howard Glasgow! He's turned

this place into a Little Shop of Horrors."

1 0

Throughout the holidays and into the new year, people from the lab and colleagues at the university would phone the Glasgows to check on Howard. Aileen would usually take the calls, because Howard was unable to remember what anybody said, and he didn't always make sense in conversation. People were really calling to say, "Hi, how're you doing? We miss you."

Then one evening JoAnn Burkolder called. When she had phoned over the previous weeks, it had been to ask about Howard's progress, but this time she was calling with good news. "I hope you're sitting down, Aileen," she said. "We've got it solved. We went over to the trailer and did a smoke test, and guess what? The hot room was venting into Howard's office instead of discharging outside. Can you believe it?"

Aileen immediately recognized that now there was a logical explanation for her husband's illness. What had happened, how it had happened, and why no one else had been affected the same way—all was clear now. Surely their financial worries were over: workmen's comp would have no choice but to approve their claim that Howard's problems were work related.

The information was especially important because, indeed, the medical specialists who were treating Howard Glasgow had not had a great deal of success in diagnosing the source of his debilitating illness. Not that he wasn't lacking in measurable medical symptoms. On the contrary. A clinical examination by Dr. Kathleen Welsh of the Division of Medical Psychology at Duke University's Memory Disorders Clinic produced results that "suggest to us that an area of the brain deep within the left temporal lobe was affected by your injury." This was an area often linked to language and memory. And a series of tests conducted by Dr. Schmechel led to a similar opinion: "Your evaluation . . . did show definite physical and neurological findings consistent with many of your symptoms and with a toxic encephalopathy." The problem was that neither examiner had been able to determine the etiology of Glasgow's symptoms with medical certainty. To date there had been no studies done on the effects on mammals of the toxin produced by *Pfiesteria*. The chemical makeup of the toxin had not even been characterized yet. In a very real sense, Howard Glasgow was a test case.

What persuaded Dr. Schmechel to believe that the dinoflagellate was the causative agent was that he had ruled out everything else, and Glasgow's condition seemed to improve when he stayed away from the laboratory. Of course, this "diagnosis by exclusion" made a reliable prognosis impossible; but in a private conversation with Aileen, after acknowledging that it would be a guess for him to say how much brain damage Glasgow had sustained, Schmechel told her she should be prepared for the possibility that her husband might regain only 80 to 85 percent of his mental faculties.

Over the next month, Howard Glasgow had his good days and his bad, but overall he became progressively better. Certain abilities came back quicker than others, and Aileen remembers vividly the day his math returned. He was laying tile with his brother, and they were trying to figure out how much material they would need, which involved calculating square footage. Aileen, listening in the next room, was thinking, *This is going to be terrible. God only knows what Howard is going to come up with*, because a week or two before, he had been unable to do basic arithmetic. She was looking around for a calculator, when she heard her husband announce the numbers he had computed in his head, and hastily Aileen punched them in. When she found, to her absolute delight, that they were correct, she had to control herself not to shout "Yes!"

But while the circuit in Glasgow's brain that was wired for math snapped back as though a switch had been flipped, his reading skills returned more slowly. He had been referred to a specialist who helped people relearn to read after they had suffered strokes, and a series of tests revealed that, among several significant cognitive problems, he had difficulty comprehending compound words. For example, he understood the meaning of the words "some" and "thing" when they were held up separately in front of him on flash cards, but when the two were joined, he struggled.

As for Howard Glasgow's take on what was happening with him throughout this period—which is to say, what he remembers thinking —even though he had acknowledged on different occasions to both Dr. Burkholder and his wife that he was having a hard time remembering things, he didn't think his problems were as great as everyone around him was making them out to be. When people told him about things he'd done—tales of uncharacteristic belligerence and irrational rages, for instance—he had a different recollection of them. To some extent he had to assume others were right, especially because so many people he trusted were saying the same things; but the truth was, at the same time he thought that maybe it was more their problem than his. And that even if they were right, they were exaggerating.

Of course, that rationale didn't explain everything. It didn't address, for example, what had gone on when a good friend of his stopped into the lab to say hello and, after a half-hour conversation, walked out the door, only to come back an hour later because he forgot something and have Howard greet him with, "Hey, good to see you. Haven't seen you in a while." Nor did it explain his poor performance on his reading comprehension tests. Presented with twenty-five sentences, each of which had a mistake—a misspelled word or a grammatical error— which he was supposed to identify within a matter of minutes, he had barely got through the third sentence when the technician said, "Time's up."

"Wait a minute," Howard had protested. "What do you mean, time's up? Nobody could do it that fast."

If his version of what was happening during this period unwittingly made a case for denial on his part, he does admit that there were times when he found himself thinking, *What's going on here? Am I one of those guys who everyone knows is crazy but he thinks it's the other way around? Is there something really wrong with me that I can't recognize?*

His recollection is very clear, however, about the discussions he had with his family when it became apparent that he was getting better and soon would have to make a decision about his future.

His mother, who lived in Cape Hatteras, and Aileen were his main confidantes, and their consensus was that given the uncertainty of what he was working with and the kinds of problems it had created, they did not want him to go back to work with *Pfiesteria*. It wasn't worth the risk, they said. He should pursue some other area of science.

Glasgow understood their concerns, and he certainly had no wish to repeat what he'd just gone through. But as he pointed out, he had three years invested in this area of research, and a lot of progress had been made. Right now, he and Dr. Burkholder knew more about this organism than anybody else in the world. So it was hard to think of giving it up and starting over someplace else.

Furthermore, he told his family, he was no quitter. If anything, he was more determined than ever to find out what exactly had happened to him. As far as he knew, he had taken the worst hit of any human being from this toxic organism, and that made this something of a personal issue. He was even more dedicated than before to deciphering its toxicological riddles, before it created problems for anyone else.

His family agreed to support whatever decision he made. So then it was a matter of waiting for Dr. Schmechel to give him a medical release to return to work.

By mid-February Glasgow seemed well enough to resume some of his duties. He was still shaky at times, and there were days when the fog rolled in. But those occasions occurred with less frequency, there were no more outbursts, and happily his neuropsychological tests were nearing the normal range. Everyone was encouraged, not least the technicians in the Memory Disorder Clinic. Aileen remembers one of their last office visits to Dr. Schmechel, when halfway through an exam a young technician stuck his head out into the waiting room, gave her the high sign, and said, "He's blowing them away." The majority of the patients they treated at the clinic were Alzheimer's sufferers, and they rarely saw much improvement, so the office was thrilled to have a success story.

When Dr. Schmechel gave Howard Glasgow permission to return to work after two months, it was with stringent restrictions. He was not, under any circumstances, to have direct association with the toxic culture. Schmechel explained that he had been "hypersensitized," and that just as some people developed allergic responses to minuscule amounts of certain poisons, he was so sensitive to this toxin that exposure to even a tiny amount could have an amplified effect.

Glasgow didn't see that as a problem. He said there was enough science to be pursued safely outside the toxic stages.

Finally, knowing about Glasgow's work habits, Schmechel forbade him to work more than forty hours a week.

When Glasgow notified Dr. Burkholder of his desire to return to work, she too laid down a set of ground rules, along the same lines as Dr. Schmechel's.

"We're going to have to figure out what we can let you do and what we can't, and you know what I'm going to say," she told him when they met in her office his first day back.

"Don't worry," he assured her. "I'm not going near the dino."

Even though the culture facility had been shut down for months, she was nevertheless glad to hear him say it with so little prompting from her.

"And I want you to work your way back gently. Don't take on any stressful activities or responsibilities."

He nodded that he understood, and that was when he told her that eager as he was to resume his normal routines, he did not feel confident enough yet to take over the lab.

Admitting that he was still not one hundred percent and that he preferred not to rush back into things was a big concession for Glasgow, who had always taken a can-do approach to whatever challenge came up. But behind his caution lay a significant realization: he knew that people would be watching him carefully for signs of persistent neurological problems. He worried that if he was having a conversation and stuttered or couldn't come up with the right word, people would automatically assume it was evidence of brain damage. And knowing how important a scientist's mind was to his reputation, he did not want to do anything to stigmatize himself.

"That's fine," Burkholder replied. "You tell me when you're ready. And please, if anything starts to be a problem, I want you to let me know."

As it turned out, one of the biggest problems he would have to confront was media interest in his case. Somehow the word had got out that a technician at North Carolina State had been hurt while working with a toxic creature in a laboratory, and calls started coming in from around the country for more information, along with requests for interviews. Although he would have preferred that no attention be directed his way, he did not want to appear to be hiding something, so he evaluated the inquiries on a case-by-case basis. Some he rejected because of the thrust of their interest, such as a TV tabloid show that wanted to cast his story along the lines of an Andromeda Strain thriller. While the *Los Angeles Times* did a more tasteful job in an article titled "Scientist Learns Toxic Algae Health Threat Firsthand," he was caught off guard by a Raleigh television station. One Saturday afternoon he was down on the coast, driving a forklift and loading sediment sand into a cement hopper as part of a sea grass project, when he heard the beating rotors of a helicopter overhead. He looked up to see a cameraman leaning out precariously, filming him for the lead story on the evening news: "Howard Glasgow is back at work."

Of all the media coverage he received, nothing upset him more than a front-page article that appeared in the *News & Observer*. It was bad enough that they played up the potential brain-damage angle. But it was the accompanying photograph that disturbed him most.

He'd had his suspicions that the photographer was striving for a dramatic effect, because he'd asked Glasgow to pose in a lab coat in front of a biohazard symbol, and he was shooting with a fish-eye lens. "Please make sure they don't make me look like a fool," he said to the writer, who assured him nothing like that was going to happen. But then she called him the night before the article was scheduled to appear: "All I can say is I'm sorry."

The photo appeared above the fold on the front page. It was in color, and it had been computer-enhanced to make him look like a zombie. His hair was frizzy, his eyes bulged in a glassy stare: he looked like a character straight out of the movie *Night of the Living Dead.*

Friends and neighbors kept phoning that morning, and essentially they all said the same thing: *Howard, are you all right? Man, you look rough.*

An old high-school girlfriend and her husband happened to be in town that day, and they stopped by the house. He knew she had seen the paper when she hesitated just as she was about to embrace him and asked, "Are you sure it's okay to hug you? I mean, I'm pregnant."

~~ ~~ ~~

Ridiculous as it was to equate *Pfiesteria* with plutonium, the fact was this organism did have the aura of something frighteningly dangerous, with powers still unknown. Even Howard Glasgow harbored a fanciful concern that went back to an experiment he had performed a year earlier: an experiment to see if *Pfiesteria* had an appetite for human blood.

Rewards in science were overwhelmingly allotted to those scientists who demonstrated originality, and one of Glasgow's strengths as a researcher was his penchant for play. Just fiddling around, he would sometimes end up conducting quasi experiments that led to unexpected results and opened up new and productive territory for inquiry.

By the spring of 1993, he had known well enough what the dino did to fish, and in a conversation with Dr. Burkholder the question had come up: Does it do the same thing to aquatic mammals? They both recalled an inexplicable die-off of whales and bottle-nosed dolphins near the coasts of New England and North Carolina in the late 1980s. Now that a series of previously unexplained fish kills had been connected to *Pfiesteria*, was it possible that the whales and dolphins that washed up had also been its victims?

Burkholder thought she knew where they could get some dolphin blood, but the people who ran marine aquariums with captive dolphins did not want any part of her research. She was offered frozen dolphin blood by one, but she had to turn it down; they had already experimented with frozen fish blood, and the dino wasn't interested. It wanted its blood fresh.

Several weeks went by, and Burkholder was still making inquiries, when Howard Glasgow found himself alone in the lab one evening, considering alternatives. Chicken blood from the supermarket was unacceptable because it had probably been frozen; the same with steak.

The next thing he knew, he was pulling a standard twenty-one-gauge syringe out of its packet and sticking it into the pad of his left thumb. After withdrawing a small amount of blood, which he dropped onto a slide, with a transfer pipette he added a slightly larger amount of dino-infested water. Then he put the preparation under a microscope and peered through the lens to see what was happening.

Three years later, almost to the day, I asked Howard Glasgow if he would mind repeating this experiment, using my blood this time. Amused, he said he would be glad to. When he handed me a needle, I jabbed my index finger sharply, smeared a bloody fingerprint on a slide, and took a seat in front of his microscope.

"What am I looking for?" I asked.

"Just watch for a minute."

I was scanning a field of pink dots and thinking the cellular world under a microscope looked a lot like a galaxy through a telescope, when suddenly space invaders struck, and they wreaked havoc.

"I really didn't think they'd go after human blood," Glasgow said, recalling what had gone through his mind that night. "I thought at most they would hunt and peck and move on."

"But?"

"See for yourself. They love it."

I thought Glasgow's use of the word "love" curiously apt, because at that moment I was tracking the movement of a single dinoflagellate, tiny and round and propelled by a long tail. It reminded me of a sperm searching out an egg as it swam among my blood cells.

The sexual aspect of its activity was sustained by the impression that the dino wasn't drawn to just any cell: it danced around to different ones—testing? tasting?—before settling on one it was attracted to.

"How does it decide what cell it likes?" I asked.

"That's a mystery," Glasgow admitted. "It could be chemosensory. It could be it goes after the older, weaker ones. Or maybe there are things it's interested in that we can't imagine. Whatever it is, we can't tell the difference, but they obviously can."

I nodded and continued to watch as the "love" analogy was carried a step further. What I had construed as a mating dance was rapidly followed by what looked like nothing less than an erotic response. After settling on a partner, the dino entered a shape-changing mode to accommodate the growth of a long and narrow appendage. In scientific jargon this is called a peduncle, but to me it looked like the dino had achieved a full erection.

I smiled—briefly. As I watched, the "male" figure mounted one of my plump blood cells, and rather than discharging orgasmically, it sucked the contents dry, leaving a shriveled, aspirated shell behind.

I whistled softly and started to rise, figuring the show was over. But Glasgow encouraged me to keep watching, so I did and saw that once was not enough for these guys. Like vampires, they moved from one cell to another, drawing blood until there was no more, then moving on to the next victim, picking up speed as they went and swelling with each consumption until they were an order of magnitude larger than when they started. That's ten times bigger. And only when they were so fat they could hardly move did they finally take a break.

It was a strange sensation, watching yourself being eaten alive. Feeling slightly queasy, I lifted my eyes from the microscope.

Glasgow and I looked at each other.

"So what did you make of this?" I asked.

Returning to his thoughts the night he experimented with his own blood, he said that although it wasn't cricket to extrapolate from humans to dolphins, the signal was strong. He had no doubt that if he were to obtain fresh dolphin blood it would do the same thing, and that this dino had a propensity for mammalian as well as piscine blood cells.

Something occurred to me at that point. The creature had consumed human blood as if it were feasting on its favorite meal after a long fast. "Is it conceivable that *Pfiesteria* has a history of attacking animals higher up the food web?"

Glasgow nodded yes.

"And not just whales and dolphins but, say, birds and animals?"

He shrugged. "Why not?"

I thought about just how high it went before asking, "When you returned to work, were you concerned that maybe it had gotten into your system?"

Glasgow hesitated, as though he wasn't sure how forthcoming he should be in answering my question. As it turned out, this was something he had never divulged before. Not to his wife, or to Dr. Burkholder, or to Dr. Schmechel. But then he hadn't been asked.

"Yes," he confessed.

In fact, it had been a question that weighed heavily on his mind in the weeks after his return, because he had heard about scientists who'd been infected by organisms in amoeboid stages that had infiltrated their bloodstreams. Amoebic dysentery was the most common infection, but he had also read about an isolate found in a nuclear power plant reservoir down in South Carolina, a real bad one that eventually attacked the brain.

It had been a week or two after he'd returned, on an evening when he was alone in the lab, that circumstances presented him with an opportunity to check his blood. Placing a sample on a slide, he put it under the lens of a microscope whose magnification was sufficient to let him see what he was looking for—and praying not to find.

In its amoebic stage, *Pfiesteria* was larger than red blood cells, so it wouldn't be hard to locate if it was there. When he did not see anything large swimming in his blood, he looked for cell destruction. Finding none, he did a count, knowing how many red blood cells and proportional white blood cells per mil he should have.

Everything looked normal, and he felt better. But he continued to check his blood in the ensuing weeks and months, and even though his fears were never confirmed, he wasn't entirely convinced there wasn't a problem that would crop up later. Just because *Pfiesteria* hadn't turned up in his bloodstream didn't mean it wasn't there. For all he knew—for all anyone knew—it could have deposited itself in one of his organs. There were all kinds of little parasitic organisms that could get into the musculature and were identifiable only from a mottling of the skin. You could have an intestinal tapeworm inside you and it wouldn't show up in a blood sample.

It is a concern that continues to haunt him to this day.

Although Dr. Burkholder had assured Howard Glasgow that his position as her lab director was secure and waiting for him when he felt up to taking charge again, she was not absolutely sure that day would ever come. Feeling as she did that she had not completely recovered from her own injuries, though she had been exposed only to a short-term, acute dosage of the dino's neurotoxin—whereas his exposure had been chronic, lasting months—she wondered how lasting the damage to Glasgow would be. When Aileen Glasgow reported the doctors' prognosis of as much as a possible 15 percent loss, Burkholder's heart ached, because even though she did not feel personally responsible for Howard's injuries, she did blame herself for not seeing sooner what was happening to him. She'd been too close to her work, she realized, too involved in the investigation, too focused on putting together a case against this spellbinding killer, to realize that her partner was crumbling beside her. The words *I should have known* played on a continuous loop in her head.

It was worse than humbling. At a certain point she decided that if Howard Glasgow did not recover, if he ended up mentally handicapped, she was going to drop the dino. Abandon the research. It just wasn't worth it.

Sooner than expected, she was given an opportunity to demonstrate just how serious she was about letting go of this project: Glasgow informed her that his workmen's compensation claim had been turned down.

Burkholder was appalled. After the smoke test revealed the faulty construction of the research trailer, she had believed that since the route of exposure was established, there would be no doubt: Howard had clearly been injured while working on the job.

"Why? On what grounds did they turn you down?"

His voice, when he answered, was weak, as though adding the fear of becoming a financial burden on his family to all the other things that had happened was pushing him toward a brink.

"When you filled out the accident report, did you say that I hadn't followed proper safety precautions?"

Oh, Christ, she thought, because yes, she had. In the process of honestly answering the questions on the form she had been required to complete as his supervisor, she had acknowledged that he had told her he did not always remember wearing a mask when working with the toxic cultures, and had not used his lab coat or gloves properly. And she had admitted noticing that he'd left fans off in the culture room and toxic tanks uncovered. But all of that was because by this time he had not been able to think clearly and logically.

She did not consider it fair to blame somebody already compromised for not following a given set of rules, especially if the person had no idea of his limitations. And she told him not to worry, she would take care of it.

Burkholder walked straight to her computer, where she composed a letter to Ms. Paula Barnes, the workmen's compensation disability coordinator at N.C. State. As a document defending a point of view about which there was a legitimate basis for disagreement, it was as powerfully argued and tightly reasoned as a legal brief. "As I attempted to state clearly in the form I submitted, and given the evidence that we uncovered this past week, Mr. Glasgow . . . was following all safety protocols, *to the best of his ability,* but . . . as a result of an insidious accident, he was apparently slowly mentally incapacitated, over what we now understand to have been at least a five-month period, by the chronic sublethal effects of long-term exposure to a neurotoxin that caused Alzheimer's-like symptoms including erratic short-term memory loss. As he became increasingly affected, he tried his best to follow the safety protocols and was unable to understand—incapable of realizing—that he was not doing so."

There followed three single-spaced pages explaining the background of the research they were doing and the frustrations they'd experienced seeking advisements about protections to be implemented when working around this organism. Then the slam-dunk conclusion: "To question his claim because the toxin with which he was trying to work safely affected his ability to do so, is analogous to concluding that, because a person lost his balance on a boat in a sudden storm, fell overboard and was hurt, he should not be granted worker's compensation because he was wearing a life jacket but did not grab for the rope (as a mandated safety protocol) as he was losing consciousness."

It was a letter she was advised not to send. When she told colleagues what she'd written, she was warned that by taking the position that her research assistant was hurt in her lab through no fault of his own, she was opening the university to a liability lawsuit by Glasgow. If that should happen, she could all but count on being sued herself because, as his supervisor, she was responsible for what went on in her lab.

She knew she was putting her career on the line by writing the letter. *But what am I supposed to do—deny it happened this way just to protect the university?* she asked herself.

She recalled her own "accident" with the defective fume hood when she was an undergraduate at Iowa State. A campus health official had told her that if she filed a claim, they were going to go after her major professor. Since she did not feel it had been his fault, she had let it drop. She had no idea what Howard Glasgow would do in this case and had no intention of asking him. All she could think was that this was the right thing to do, and she mailed the letter. Then she called Glasgow to let him know she was sending him a copy, and if need be she was willing to go to court and testify to everything she'd written.

While awaiting the outcome, she busied herself with the myriad other matters that demanded her attention. When she confronted the contractor of her laboratory facility and informed him there was a serious airflow malfunction in the design of the ventilation system, he denied any knowledge that the facility would be used as a lab and claimed that the university had approved the plans and that he had understood it would be used "for business occupancy"—this, even though his own floor plans were stamped "Fish Lab" because he'd had difficulty spelling dinoflagellate. After passing this information to the university, which was contemplating legal action against the contractor, she took part in the Health and Safety Center's assessment of the biohazard threat posed by this dinoflagellate. A thorough inspection led officials to decide that because engineering controls for personal protection were based on the toxicity of a particular agent, and no one knew what they were dealing with, if Dr. Burkholder reopened her lab it was going to have to be a Level 3 biohazard facility. This was a system reserved for handling pathogenic and potentially lethal agents, in which all procedures were conducted within physical containment devices by specially trained personnel wearing protective clothing.

When she thought about how much time this was going to take to put in place, Burkholder's morale, already low, dipped deeper. By her calculations she had lost almost half a year of research time: the three months her lab had been padlocked since Howard had gone on sick leave and, prior to that, three months of ongoing experiments that had to be shut down. Personnel time, supply moneys, material they were poised to work with—all had gone down the drain. This would be a major blow to any research program, but adding in the special circumstances operating here—a breakthrough discovery that other scientists coveted and had begun taking steps to control—she knew it had the potential for disaster.

Since she was unable to do any new work in her lab, she spent most of her time during this period writing furiously: grant proposals that would allow her to make up for what she'd lost, and scientific papers that covered the advancements in her understanding since the *Nature* letter had been published. In a way, the concentration writing required served as an escape from the worries nagging at her, but from time to time as she sat at her computer, her mind would wander. And when it did, it usually returned to a conversation she'd had in early December at a biotechnology workshop in Wilmington, North Carolina. She and Dr. Peter Cover, a colleague of Dr. Noga's at the veterinary college, had been standing in the cafeteria with lunch trays in hand, and he said he'd heard that Howard Glasgow had been sick and asked her what happened. After she told him what she knew, he asked her what symptoms Glasgow was exhibiting. She described them, and Dr. Cover asked if she'd shared this information with Noga. When she told him no, they were barely on speaking terms, he said to her, "JoAnn, you need to speak to Ed."

Something in the way he had spoken implied that there was more he could say. "Are you trying to tell me something, Peter?"

Dr. Cover held her gaze. "If you ever repeat what I'm about to tell you, I'll deny it," he said. "I'll swear I never said it and that you're lying. But I am absolutely certain Ed Noga knows a lot more about the effects of this dino than he has let on. And if he had shared some of that information with you, it would have prevented what happened to Howard."

～～ ～～ ～～

People confided in Peter Cover. Graduate students at the veterinary college, his own as well as those with a different major adviser, would come to him with their problems. When he thought about why people approached him, he figured maybe it was because they knew he was familiar with the intrigues and frustrations that were part and parcel of academic life. But it also could have been because he ran an open lab in which students, assistants, and professors worked together with a tight sense of collaboration.

His ability to make his students feel they were part of a team was in marked contrast to the philosophy and style of Dr. Noga, who had a reputation around the vet school as a secretive researcher and a "control freak." It was widely known that Noga liked to compartmentalize, parceling out work while withholding background that would grant lab workers an overall sense of their project. And once they turned over their data, no one really knew what he did with it until he got around to publishing it, which was always in his own time and on his own terms.

While there was nothing inherently wrong with this approach to research, it had, in Cover's opinion, created certain problems for Dr. Noga. The papers he had published over the years were fairly well respected, but Noga himself had become the target of a certain amount of criticism. On several occasions he had been accused of taking credit for work done by his graduate students, not putting them on a paper as coauthor, or refusing to let them review before publication a manuscript on which their names appeared.

Though it was hard to say whether these criticisms were the result of collecting data piecemeal and analyzing it on his own and of his secretive nature, what could be said was that Dr. Noga's interpretations sometimes exceeded the boundaries ordinary scientists would have drawn. His way of holding his cards close to his vest figured into why some of the faculty at the veterinary college refused to work with him; and all of this explained why many of his own students went to Dr. Cover when they had something on their mind.

Stephen Smith, Dr. Noga's postgraduate student who discovered that a dinoflagellate was killing fish in their aquariums, first approached Dr. Cover in the summer of 1989, when he was experiencing disturbing physical and neurological symptoms of no known origin. He pointed to skin lesions that wouldn't heal and talked about a persistent tingling sensation in his hands and feet and difficulty in maintaining his coordination when walking. Neither Smith nor Cover made an association with the dinoflagellate. In fact, Smith was also doing experiments with a polychromite gel that was known to be a neurotoxin, and for a while that confused the issue.

In retrospect it would be painfully clear: this was a time when he was working around aquariums filled with the dino in a closed room, and the tanks were so hot that he could put a new fish in and it would start showing neurological symptoms in five minutes and be dead within fifteen. It never crossed Smith's mind that it might not be a good idea for him to be reaching into the tanks with his bare hands and picking out the dead fish before adding new ones. Besides, in Dr. Noga's lab, *he* was the man in charge, and if there was something you should know, *he* would tell you.

When Stephen Smith reported his symptoms, Dr. Cover did not have any answers, but he did maintain a concern about Smith's health and stayed in close touch with him as his condition deteriorated. A series of medical examinations produced different diagnoses, including Lyme disease, autoimmune peripheral neuropathy, and multiple sclerosis. When further tests ruled out each possibility, the doctors were so baffled that there came a time when Smith found himself lying on an examining table while two neurologists sat on the floor flipping through textbooks and a third stuck him with needles.

In the end, everything else having been eliminated as a potential cause, the medical specialists decided it had to be the dinoflagellate. And their suspicions were supported when, shortly after Smith ceased working around the dino, his condition began to improve, although he did continue to experience problems with concentration for several months.

When an appropriate situation presented itself, Cover spoke to Dr. Noga about Stephen Smith's illness, asking Noga if he had considered the possibility that it was connected to the toxin produced by the dinoflagellate. Not one to welcome scrutiny, and even less accommodating toward recommendations that implied criticism of the way he ran his lab, Dr. Noga said he did not believe that Smith's problems had anything to do with the dino. He was emphatic that there was nothing in the dinoflagellate literature that would explain Smith's problems, which he dismissed as an "idiosyncratic" reaction.

Dr. Cover found the whole situation astounding, not least because he would have handled it so differently. He was a cautious researcher who maintained a close watch over safety in his lab, because he had seen a major professor undergo a nasty lawsuit. During Cover's training, in a lab that worked with infectious agents, the department head was sued when a janitor claimed he had come down with a disease from a chimpanzee he'd been asked to dispose of. The complaint ultimately failed—it was demonstrated through safety records that everything left over from the experiments was sterilized and there was no way the janitor could have been infected—but the experience was sobering, a cautionary tale for any researcher working with hazardous agents.

At the time, Dr. Cover did nothing more than have a conversation with Dr. Noga, because Noga was right: the case against this organism as a threat to human health was incomplete. Cover also knew that even though he did things differently, the phenomenon was not uncommon within academia and research universities across the country: ambitious researchers held information close, letting their technicians know only what they needed to know to get the work done. Students and assistants were sometimes asked to do work that made them potentially vulnerable to health problems, and they went ahead, trusting that their adviser knew what was safe. Down the road, health and safety issues were raised, but they were usually dismissed or denied. This was almost as much a part of the scientific process as trial and error.

Almost two years passed before Cover was approached again about the way Dr. Noga was handling his research with the dino, and this time it was one of Noga's graduate students, Janice Kishiyama, a young woman of Japanese ancestry. Dr. Noga had, typically, not provided any history but simply given Kishiyama a collection of tissue samples from the skin, gills, liver, kidney, and brain of fish that had been exposed to and killed by the toxic dinoflagellate, telling her to examine and process them for histopathology.

What Kishiyama saw, she would later say, "scared the living daylights out of me." The samples, which were fixed in formalin and pressed under glass slides, looked as though they had been hit by a powerful ray that eroded skin, causing massive ulcerations, and destroyed the integrity of internal organs by wiping out the immune system—all in an extremely short period of time after exposure. She was so alarmed that she immediately typed up the results and gave Dr. Noga a print-out, along with an expression of her concerns.

"Based on the harmful effects this toxin is having on fish flesh," she said, "aren't there things that people working with it should be told to watch out for?"

Dr. Noga had given her a stern look that she interpreted to mean: *If I wanted your advice I would ask for it.* But she had not thought the question out of line. Indeed, she believed that the test results sent up red flags that any reasonable researcher would recognize; and any responsible one would immediately issue a warning to everyone who potentially could come in contact with this dinoflagellate in a laboratory setting: *There very well could be adverse health implications for humans exposed to this dinoflagellate. You should have no physical contact with it.*

Given the immune system compromises the fish were also showing, she felt that warning should also say: *The white blood cells in fish are disappearing as well. There is still more to be learned, but in the meantime you should be aware of this and take appropriate precautions.*

After Kishiyama expressed her belief that they should make this information available, Dr. Noga let her know that the lab was his and he did not plan to make the findings public at that time. And as she would say later, "With Ed Noga, you can suggest things and that's it. What he does with what you say to him is outside your control. I would have thought that at the very least he would have taken a step back, looked more closely at what he was dealing with, and found out what we, as humans, should be aware of before proceeding. Particularly when he was asking people in his own lab to work around it and knew that people in other labs were handling similar material. I don't know why he didn't do anything."

That wasn't entirely true. After Kishiyama finished reading out the slide samples, Noga moved her to other experiments and never allowed her to work with the dino again.

These events had taken place in the midsummer of 1992, almost a year and a half before Howard Glasgow went down; but it was not until the fall of 1993, several months before Glasgow's symptoms, that Kishiyama, who had said nothing to anyone out of fear that if she spoke out Dr. Noga would not allow her to graduate, finally confided her fears to Dr. Cover in a private conversation.

At the time he was given this information, Cover had done what he usually did when graduate students complained to him about their major professors: he referred them to a chapter about petty tyrants in one of Carlos Castaneda's books. Petty tyrants, according to Castaneda, were basically people who defined themselves in terms of the control they wielded over others' lives, and their need for control usually emerged from a deep personality flaw. The best way to handle these people was not to fight them, because they throve on resistance. Rather, you should take a tolerant approach, observing them and learning about their weaknesses, and in the end, Castaneda's mentor, Don Juan, predicted, these people would eventually self-destruct.

But when Dr. Cover heard about Howard Glasgow, and he put that together with Stephen Smith's health problems and Janice Kishiyama's concerns, the pattern that emerged was so bothersome that Cover was uncomfortable with the advice he had given. It was one thing to keep a lid on data as it related to sensitive areas of research for publication purposes or in order to compete more effectively for grant moneys.

It was another matter entirely to withhold from other researchers information that would enlighten them to known or suspected hazards. To Cover's way of thinking, safety concerns should come before anything else.

Peter Cover found himself in a quandary. He did not think that Dr. Noga would deliberately put others in jeopardy. Yes, Noga was a driven scientist, driven to the point that he had few hobbies, few friends—nothing but his work. But Cover continued to believe that his colleague's behavior was merely a reflection of the research culture in academia that maintained you didn't share data because it might allow a competitor to catch up or pull ahead.

In the meantime, however, human beings were suffering who very possibly could have been spared if just some of this information had been passed along. And this, in Cover's mind, created an obligation to alert others to the potential health risks. Which is why he said to Dr. Burkholder, "JoAnn, you need to talk to Ed."

As for why he added that he would deny his conversation with her if she told anyone what he'd said, that related to another complicating factor. He was sitting on the Institutional Animal Care and Use Committee, and each member had been required to sign an agreement of confidentiality. It was the only way they could get researchers—some of whom were dealing with trade secrets, patentable techniques, intellectual properties—to talk openly. Committee members had to be able to say, "In order for us to function, we have to have access to information so we can make proper decisions. Thus anything you tell us will be kept in the strictest confidence." And because he felt his discussions with Kishiyama—at least in those parts that dealt with studies that were being done on animals and what safety concerns were or were not being addressed by laboratory workers—were conducted by him in the context of his capacity as a member of the committee, Dr. Cover, conflicted over what was right and wrong in this situation, what was protected and what was not, was indirect in the action he took: unofficially nudging Dr. Burkholder in a direction that would lead her to the same information he had, yet covering himself by saying he would deny their conversation.

As events progressed, Cover's situation became even more difficult. Feeling as he did that the issues raised were legitimate areas of discussion for the Institutional Animal Care and Use Committee, he brought them up at the next meeting. Since most of the committee was composed of faculty members who had to confront safety issues within their own labs, he expected them to share his concerns and recommend a full inquiry. But the reaction of his fellow members surprised him. Most of them got stuck on the fact that there was still no absolute scientific proof that this organism endangered human health. No clear and measurable cause-and-effect relationship had been established yet, they pointed out. Different causes could explain what happened to Smith and Glasgow, he would remember someone saying. It could be that Smith was allergic to something in the water and Glasgow was having a reaction to something on the fish.

It was about this time that Dr. Cover learned that as a result of Howard Glasgow's injuries, the Health and Safety people on campus were initiating an investigation. And included in their inquiry was an effort to substantiate a rumor about Dr. Noga that had investigators asking, "What did he know? When did he know it? What did he do with it?"

Keeping a personal distance from the investigation, Dr. Cover followed its turns as closely as he could without compromising his role as a disinterested party. And what he witnessed convinced him that this inquiry was bound for failure. In particular he would remember a conversation he had with one of the investigators about Noga. Acknowledging that he was trying to chase down rumors about Dr. Noga, the investigator said that Noga continued to maintain that no one to his knowledge had been hurt in his laboratory by this dino.

"I think he's holding back," the investigator admitted, "but I don't know enough about what's going on to be able to ask the good questions. When I query him, either he denies everything or his answers are brief and give me nothing that could lead me closer to the point I'm getting at."

Dr. Cover had been incredulous. Though he held his tongue, he wanted to say: *Hey, it's your job to ask the right questions. And to put the pressure on when you don't get them.*

He didn't know why the investigator was so tentative. Was it lack of competence or did the man believe there really was no problem?

But if Cover had been asked to bet, he'd have put his money on a different explanation. He believed that this investigation was not truly geared toward getting to the bottom of this situation. No one wanted to know what had gone on, because if the investigation turned up the fact that a tenured professor had knowingly withheld information that jeopardized the health of his students and colleagues, not only would it result in lawsuits but the investigation would surely broaden to include other researchers, and the upshot would be a scandal that would embarrass everybody. This, he felt, was what accounted for a passive response to the health and safety issues raised by this case. Rather than take a proactive approach to ferreting out the truth, particularly when the potential ramifications were so unpleasant, the university was going to look the other way, unless it had no choice but to face the situation.

By nature, Cover was not a whistle blower. He was a teacher, someone who moved people by persuasion. But when he realized that the university wasn't taking this situation seriously, was doing just enough to be able to give the impression that it had looked into the matter, he went to the committee files and printed out the minutes of the discussions his committee had had concerning this affair, and he sent a copy to Stephen Smith. He had remained in touch with Smith—who was now teaching at Virginia Tech University—and knew that he was continuing to suffer both mentally and physically from effects he attributed to his exposure. Cover's concern was that with some marine toxins the effects not only were irreversible, they could progress with time. He knew Stephen Smith had a family—a wife and two daughters —and it worried him to think that the man's neurological problems might worsen in the years to come, affecting his ability to provide an income.

"Look, Steve," Cover said, "you want to be friends with everyone, even someone who might be responsible for injuries you've sustained, fine. But if someday you should decide you need to do something about it, I want you to have this for your records. It documents that university officials were made aware by our committee of our concerns and suspicions. Do whatever you want with it. I'm not going to take any further action. But if asked in a court of law what was going on, I'll tell the truth to the best of my recollection."

After that, Dr. Cover did not discuss the matter with anyone—until he spoke with me.

When I asked Dr. Noga about these matters, he acknowledged that science was an "ego-driven profession" and that the desire to find answers did tend to blind one's judgment. He also admitted that the "only thing that is really worth anything [in science] is intellectual property," and that it was important, if you were motivated by peer recognition and all the things that accompanied that, "to protect your ideas." But he denied that any of this factored into his behavior regarding this discovery.

Dr. Noga also denied that anyone working with this organism in his lab had ever been adversely affected. When asked for his explanation of the neurological problems experienced by Stephen Smith, while saying he wasn't a neurologist and therefore was unqualified to comment specifically on clinical symptoms, he went on to describe the room in which Smith had been working as lacking ventilation, and offered a build-up of carbon dioxide as a possible explanation. And he stated that the first time he realized its toxic potential to humans was after Howard Glasgow was hurt, assuring me that if this information had been brought to his attention at an earlier date, he would have immediately shut down his lab.

By the middle of March, JoAnn Burkholder's lab was in chaos. Without more attention and supervision than she had the time to give, many of the eight or nine students and lab assistants who worked there would come in around ten or eleven, take long lunches, and leave at three. They more or less knew what was expected of them, but nobody was doing more than it took to get by. Burkholder, who realized she was going to have to do something about it, had been waiting for Howard Glasgow to let her know when he felt up to taking charge again. Having told him to take his time and not rush things, she didn't want to pressure him to come back sooner than he was ready to, but she didn't know how much longer she could wait.

When he finally came to see her one afternoon, he looked like his old self, in fine spirits after having learned that her letter had apparently worked and workmen's comp would honor his claim. But she could tell that something was troubling him, and she soon learned why. "I don't like what's going on in the lab," he said.

"You mean nobody is working very much and they're getting away with things they shouldn't be getting away with?"

"Yeah," he said. "Exactly."

She waited a beat. When he did not continue, she said, "But you're not ready to take over yet, is that it?"

Glasgow looked down at the floor for a long time, and then he looked back up and said, "Give me a few days."

She understood it was going to be difficult for him to reposition himself as lab director. After all, before he'd left he had alienated himself from many of the students, and not withstanding the reason, a certain amount of respect had been lost. Even now, people weren't sure how much he had recovered, because, as he admitted, he was still having trouble reading.

A week later, Glasgow knocked at her office door. "I think I'm ready," he said.

She gave him a long look. "Ready for what?"

He returned her look. "The lab is a mess, and it's driving me crazy."

It was what she had been waiting for him to say. "Well, what do you think we should do about it?" She stressed "we."

Their eyes locked, and an energy was exchanged that she never thought she would experience with him again.

His expression softened. "I'd like you to write a letter and put it in the lab, saying that I am reassuming responsibility as lab director from this point on, and everyone should report to me."

"I would be absolutely thrilled to do that," Dr. Burkholder said, and she posted the notice that day.

Howard Glasgow had not gotten around to reporting his after-hours discovery that *Pfiesteria* had an appetite for human blood before Dr. Burkholder showed up in the basement laboratory on some other matter. Offhandedly, he mentioned that she ought to check out what he had under the microscope.

Put any number of algae under a scope, and she could identify them as to genus and species. But under magnification, she did not recognize human blood.

"What's this?" she asked, dialing in the focus.

"Just watch it for a minute."

She watched, and saw the dinoflagellate darting around and aspirating blood cells, an activity she was familiar with by now. "So what am I doing here? What's going on?"

"A funny thing happened the other night," Glasgow began. "I was taking a needle out of the syringe packet . . . and you know how you have to pull the plastic back?"

Her head lifted. She knew him well enough by now to know this was a prelude to something startling.

"Howard, what are you trying to tell me?"

"It pricked my finger."

His sheepish grin told her the rest of the story.

"And from there your blood bounced perfectly into a petri dish. And it was a petri dish the dinoflagellate just happened to be in. Is that what you're saying? Are you telling me this is your blood?"

He gave her a smile and a shrug. "What was I supposed to do? Let it go to waste?"

Burkholder shook her head. "I'm going to put an all-points bulletin out to the lab: 'Watch out for Howard Glasgow! He's turned this place into a Little Shop of Horrors.' " But she was laughing too, because even though he'd gotten ahead of the research plan, she knew that when he dabbled, his findings could be very interesting.

Immediately she wanted him to repeat what he'd done, giving all the details, from the beginning.

When she had considered a marine mammal study, it had not occurred to Burkholder that the dino might cue in on human blood, so her amazement could not have been greater. Wherever there was blood in the petri dish, the dinos dove toward it, sometimes fighting each other over a single cell. Watching this creature devour fresh human blood was a "surreal experience," she would remember. "I just stood there with my mouth open."

Of course she couldn't let the investigation rest there, and over the next few weeks they tried to expand their understanding of the dino's insatiable craving. They tested for grazing rates—how much it ate, how fast it ate, and how that compared with fish. They found that each

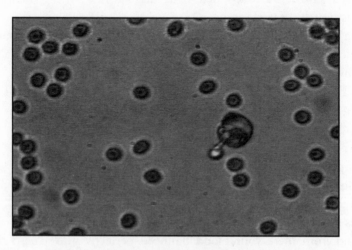

Pfiesteria piscicida **feeding on a human red blood cell.**
Photo by Howard Glasgow.

dino could consume five to eight human corpuscles in less than ten minutes, before it got so sated it could no longer swim. This was bad news. It was about the same as for fish blood.

On the other hand, there was some good news. Yes, fed human blood, the dino gorged itself. But given unlimited opportunity, it ate itself to death. Without fail, three or four days after splurging on human blood, every dino in the petri dish was dead.

Notwithstanding the fact that *Pfiesteria* apparently could not survive on human blood, Burkholder and Glasgow each realized that adding this dimension to the dinoflagellate raised serious human health questions. Glasgow found himself imagining a scenario in which a commercial fisherman, out trawling for menhaden or shrimp in one of North Carolina's estuaries or sounds, unknowingly came across a *Pfiesteria* bloom. And while working his nets, he accidentally cut himself. Thinking nothing of it, he washed the wound off in water he didn't realize was crawling with the microscopic equivalent of piranha.

Burkholder's thinking picked up where Glasgow's left off. Although she too could envision adverse effects on those who came in direct contact with the dino, she was just as concerned about the potential hazard from inhalation of aerosols containing the toxin to both occupational users of the water and recreational users: swimmers, windsurfers, water-skiers, boating enthusiasts, and amateur fishermen.

Furthermore, while she was cheered by the news the dino did not thrive on human blood, she wasn't exactly thrilled, because a subsequent experiment revealed that it liked white blood cells as well as red. White blood cells were the body's first line of defense for fighting off infectious diseases. Lower the number of white blood cells, and you became more susceptible to bacterial and viral infections. If immune system suppression was a by-product of exposure to this organism and its toxin, you created the potential for increased vulnerability to a range of opportunistic diseases. This was what happened when a person tested positive for HIV—an analogy that occurred to Burkholder because of Dr. Noga's earlier description of the immune-suppression syndrome in fish as "fish AIDS."

These were intriguing issues. But at the time, Burkholder felt she had no choice but to file them for later consideration, because she was already having to fight to get the state of North Carolina to believe this organism killed fish. She could only imagine what her opponents would say if she were to announce it also had a vampire's appetite for human blood.

~ ~ ~

Her fight with the state began with the report she wrote for the Albemarle-Pamlico Estuarine Study group in 1993. When she'd been awarded a grant, the group's director, Randy Waite, a former EPA official, told her he was desperate for an answer to what was killing fish in North Carolina's estuaries. And how and why. He complained that each time he'd gone back to his underwriters for more money to continue the study, they would ask him if he had identified a causative agent yet. "We're getting close," he said he told them. But the truth was that after dispensing millions of research dollars to scientists around the state, all they had come up with was requests for more money. He said he was hoping that she would be able to help them identify the cause of the mortality at last.

What he had not brought up with her was the controversy that surrounded the entire APES program and was sure to engulf any researcher who came up with definitive findings. About that she found out on her own.

Unknowingly, when she accepted the APES money, JoAnn Burkholder had been thrust into the middle of a management debate that had been going on for at least twenty years. Essentially it boiled down to two opposing points of view regarding what was killing fish in North Carolina's estuaries. Lined up on one side were those who said it was a product of the natural system. Sheltered by a long ribbon of coastal islands known as the Outer Banks, North Carolina's sounds and estuaries were shallow, wide, slow-moving bodies of water. They qualified as lakes, really, and flushed only occasionally. The argument went that during the summer months, fresh and salt water, which didn't mix well, would form layers, with the salt water sinking to the bottom. This created a condition known as a "salt wedge," where the level of dissolved oxygen was so low that it severely stressed fish. At the same time, the hot summer temperatures would promote the growth of algae, and when these blooms died, the large amounts of oxygen used by bacterial decomposers would further deplete oxygen levels. The combined result? Fish suffocated.

Those on the anthropogenic side of the argument, while not denying that happened, said the coup de grâce was delivered by the phenomenon known as eutrophication, the discharge by towns, farms, and industries of pollutants into the rivers. What was really killing fish were the cumulative effects of dirty water.

When she initially received funding from APES, Burkholder had been vaguely aware of this dispute, but she had avoided attending to the issues because she had not wanted to get involved in a public conflict. In that regard she was typical of scientists who were taught in graduate school to adhere to the paradigm of the disinterested investigator committed to the pursuit of knowledge for its intrinsic merits. Perhaps good would come from your work, you were told, but it was not your place to become an advocate of your science or involved in its application in the real world. Besides, the last thing Burkholder wanted was to be drawn into a controversy. It tended to scare off potential funding agencies, and she was having enough trouble raising money for her work.

And so she had gone about the collection of data without regard to the implications—sampling water taken at fish kill sites for the presence of this organism, analyzing the milieus in which it seemed to thrive, and setting up control studies in different locations. Then conducting experiments—called bioassays—in the lab in an attempt to duplicate the findings.

Since the age of sixteen, when she first read about the plight of the Great Lakes, this was what had interested JoAnn Burkholder: the study of aquatic life, especially algae and how they responded to environmental variables. Until this time, most of her work had been in lakes, which were relatively easy to study, and rivers, which were more difficult because the water flowed and there was more turnover. Estuaries were the ultimate challenge. You had the ocean coming in and the rivers flowing out, increasing the complexity of the system by an order of magnitude and making it harder to control conditions and make predictions.

Of course JoAnn Burkholder believed, as did everyone informed about coastal water issues, that the single biggest threat to the well-being of an estuary was pollution. Most people, thinking pollution, pictured an oil spill. But the more pervasive, more insidious, more devastating threat to the health of most waterways, especially in coastal bays, was excessively high concentrations of nutrients—nitrogen and phosphorus, primarily—that came from point sources, such as municipal sewage systems that discharged treated wastewater into the rivers, and non-point sources, in the form of pesticides and fertilizer that washed off farmlands during heavy rainfalls. They were called nutrients because in small quantities nitrogen and phosphorus were beneficial to aquatic ecosystems, nourishing the growth of sea grass and algae and microscopic animals that made up the base of the food web. But uncontrolled, runaway nutrients could shift the ecological balance in an unhealthy direction.

The way Burkholder explained the deleterious effects of nutrients when she was lecturing in class was to say, "If you go out and fertilize your houseplants or your garden, what do you do? Do you take a bag of inorganic nutrient salts and pour it on the ground?" Of course her students always said, "No, you add water," to which she would respond, "Why do you do that?" "Because water makes the nutrients more soluble, more accessible, so plants can take them up." "Right," she would say. "Now picture plants that have a whole lot of surface area floating in a big body of water, which is what you have in estuaries with algae. They are sucking up nutrients that are already dissolved and highly accessible, and it becomes a case of too much of a good thing too fast. It causes the whole community to shift. The normal phytoplankton that are there are outcompeted by undesirables who like a heavy load of nutrients. Sometimes even toxic species that previously existed in small and inconsequential numbers will be stimulated by this inadvertent environmental manipulation."

Knowledge of the impact nutrient-rich waste had on ailing coastal waters did not originate with Dr. Burkholder. The information had been around for decades. What she brought to the discussion with her research for APES was a technical database, gathered over a period of two years both in the field and in culture, that showed a direct connection between nutrient enrichment and the presence, the population size, and the toxic activity of *Pfiesteria*. The creature reproduced like crazy in water containing high levels of phosphorus and nitrogen.

Furthermore, she was able to pin at least one-third of the major fish kills in the Albemarle-Pamlico estuarine system in the years 1991 and 1992 on this same organism. And because of the difficulties she'd had reaching fish kills in time to obtain water samples while fish were still dying, she assumed the percentage was underestimated.

Burkholder had found that one of *Pfiesteria*'s favorite haunts was the Pamlico River. Evidence collected by the state's own biologists confirmed that approximately 70 percent of the fish kills linked to the organism had occurred in phosphorus-enriched areas on the Pamlico —some near sewage outfalls where municipal wastewater was discharged into the river, but the majority near where Texasgulf, Inc., a multinational corporation, operated the largest open-pit phosphate mine in the world, on the southern bank of the Pamlico. After phosphate ore was dragged from the earth with enormous cranes, mixed with water, and pumped to a mill, the waste produced by the phosphate-making operation—thousands of pounds of phosphate dust each day—was dumped back into the Pamlico. Texasgulf accounted for almost 50 percent of the total phosphorus in the river, and it had been allowed to do this for almost thirty years, because the state claimed it could not produce hard scientific evidence that the discharges were harmful.

It did not surprise Burkholder that the biologist Kevin Miller, who had sent her samples from a fish kill that allowed her to identify the organism in the wild, had found a hot spot right around the point from Texasgulf. The conditions the company created in the Pamlico were perfect for *Pfiesteria*.

The whole idea behind APES had been to produce a document that linked basic science research to a set of recommendations that would be presented to the North Carolina Division of Environmental Management, to inform and guide management practices throughout the state in the coming years. So naturally Burkholder thought she was giving what was asked for when, at the end of the preliminary report she submitted to Randy Waite in 1993, she wrote: "Based on the available data, we recommend that the potential for stimulation . . . by phosphate enrichment be seriously considered by NC DEM in its ongoing efforts to develop improved guidelines for protecting North Carolina's estuarine water quality . . . [and] further reductions by existing point and non-point sources should be considered by NC DEM as a long-term strategy in strengthening protection of estuarine waters and fishery resources."

The final report for APES was to be a peer-reviewed publication, so a draft of her paper was circulated among outside scientists for comment. This did not concern her; the work was verifiable, and her recommendations were not drastic. All she was saying was that in order to reduce the number of fish kills caused by *Pfiesteria*, something would have to be done, something as simple as slowing the flow of nutrients into the estuary.

She was stunned by the review that came back from the Division of Environmental Management. Her information on nutrient enrichment and her conclusion that some fish kills were linked to this dinoflagellate were "specious," it said, because they were based only "on laboratory studies and two years of field data." Regarding her recommendations: "Until stronger evidence links [this dinoflagellate's] deviant behavior to causes that can be controlled through the management of point and non-point pollution, we feel that the implementation of management strategies specific to the organism would be premature."

Dr. Burkholder had had no idea how governmental agencies went about incorporating pertinent scientific information into their policy-making process. "I'd never even thought about it," she would say later. But given the high level of public concern about the deteriorating health of North Carolina's estuaries and fisheries, she had assumed that her recommendations would be welcomed. By identifying a causative agent and the environmental conditions in which it flourished, she was supplying state environmental agencies with a factual basis on which to carry out their mandate to protect a natural resource under their jurisdiction.

"I thought they were out to protect water quality in the estuaries and that they would be able to use information about a nutritional trigger because it would give them ammunition when it came to trying to develop better regulations and guidelines in terms of nutrient loading. And what were the two parts of the report they took issue with?

The two central findings: the fish kill information and the nutrient connection."

So what's left in the report that you do believe, folks? she was left thinking.

If it had been a matter of DEM locating a soft spot in her hypothesis, or if DEM had said that after considering the evidence in depth it had arrived at conclusions very different from hers, that would have been acceptable, because it would have set terms for discussion. But that wasn't how she perceived this rebuttal. Rather than questioning the validity of her studies, the DEM critique, as she interpreted it, called into question her reliability as a scientist.

On the face of it, DEM's response would appear to be unreasonable. JoAnn Burkholder had received both national recognition and acceptance from the international science community for her identification of *Pfiesteria* as a fish pathogen of consequence. But apparently that wasn't good enough for the state's Division of Environmental Management, which, by this action, seemed to be slighting her reputation.

She picked up the phone and called the director of APES. "What is going on here?" she demanded to know.

Randy Waite groaned. "That's DEM, JoAnn. Their role in this study has been to give us nothing but trouble. They say they're just trying to weed out fact from emotions on environmental issues, but they use that excuse to block just about everything we try to do. It's best just to try and ignore them."

"But this is ridiculous," she shot back. "It would be irresponsible for these people to act as though this information doesn't exist."

"I know, I know."

As the situation was explained to her, from the outside it appeared that the Division of Environmental Management had been unable to strike an equitable balance between the development and the conservation of North Carolina's natural resources. And while it maintained a public posture of wanting to have a scientific basis for making sound management decisions, what often happened was that it demanded such a high burden of proof from scientists before it would enact a regulation that nothing got done. Whereas the Environmental Protection Agency was willing to act when it was 90 percent sure, DEM wanted 100 percent proof. And when officials were asked why it insisted on this level of certainty, the standard answer was that they did not want to move ahead on costly regulations when there was still some doubt in the air.

Linking fish kills to pollution and recommending phosphorus reductions was just the kind of information APES was looking for, Waite told her. But it was not what DEM wanted to hear.

Burkholder fumed silently. She knew that in science, particularly in the environmental field, there was no such thing as absolute certainty. You could never prove causality 100 percent of the time. The best you could do was amass evidence from many different directions and strongly correlate it. That was why, in her report, she had written that *Pfiesteria* was "implicated" and nutrients were "linked." The problem with those words was that people looking to challenge your conclusions would focus on their indefiniteness. And if you'd done your work in the field, they'd say it needed to be repeated in the laboratory. And if you did it in the lab, those were called artificial conditions. It was the way "scientific truth" could be exploited and abused.

"Who do I talk to over there?" she asked.

Jimmie Overton was the man whose name she was given, and it surprised her. She was acquainted with him: not well—several years before, he had asked her to look at some water samples for the Water Quality Section of the division—but well enough to know he was an outdoorsman who liked to hunt and fish, and to feel she could go straight to the point when she got him on the line.

"Have you read the review of my report?"

Burkholder's outspokenness, combined with her command of her subject, could be intimidating, and Overton was clearly caught off guard.

"Well, actually, I haven't had a chance to look it over closely yet, JoAnn. Why don't I get back to you?"

"Fine. But in the meantime you can tell your folks that I don't accept the central premise of their review."

"What central premise?"

"Saying this dinoflagellate doesn't kill fish. Saying I shouldn't be saying such things because I only have two years of data."

"But JoAnn, that's the way things are done here. We'll need to see ten years of data before DEM will be willing to consider that this dinoflagellate might kill fish."

She thought about that. Thought how at this point the prudent thing would be to say, *Okay, I'll be a good scientist and come back in ten years.* But she just couldn't bring herself to do that. This was what she had found in the estuaries, and if nothing was done to reverse present trends, she believed it was only going to get worse.

She felt she had no choice. "I'm sorry," she said, "but I think that's absurd. I may only have two years of data, but I'm standing behind it. I'm not backing off anything I've written. And I'll bet you your next deer-hunting season I'll be back with proof a lot sooner than in ten years."

~~ ~~ ~~

Her experience with the APES report a year earlier explained JoAnn Burkholder's reluctance to come forward with information about Glasgow's impromptu experiment when she first learned about it. That didn't mean she intended to drop the matter, however. On the contrary, she decided she would have to develop a database on her own, so that when she did come forward she wouldn't be laughed at.

Two considerations influenced her thinking at this point:

She knew one had to be careful about generalizing across multiple species. Animals' organ systems differed in their analogies to the human system, and without knowing the toxin the dinoflagellate produced or its biochemistry, and without knowing exactly how it was transmitted to make people sick—she had reason to believe you could be hurt through inhalation and dermal exposure, but she did not know what happened if you ate fish infected with the toxin—she wanted to be careful about going out on a scientific limb.

At the same time, she was aware that in some circumstances the prudent response involved making assumptions. Perhaps you could not prove that just because fish exposed to this toxin under the right conditions would die, a cumulative exposure would also be fatal to human beings; still, if you observed fish going through the natural course of a disease that ultimately had a bad outcome, and you saw human beings undergo the first couple of stages of a similar-looking disease after exposure, you certainly had a right to suspect that they might take a similar path over time.

Coupled with that thought was a question: she knew what she and Glasgow had sustained firsthand in dealing with the toxic dinoflagellate in a closed laboratory setting, but was something similar taking place in an environmental setting?

At this stage she kept what she was thinking to herself, but she began to make discreet inquiries. The people she especially wanted to talk with were those most likely to have been exposed repeatedly and for long periods of time to *Pfiesteria* blooms: commercial fishermen. The problem was, fishermen were notoriously closemouthed. They didn't like to discuss where they fished, what gear they used, or the size of their catches. And she had already run into opposition from commercial fishermen who were concerned that her work with *Pfiesteria* was generating negative press about North Carolina's fisheries.

Quietly she began to make phone calls, starting with fishermen she had met while taking samples on the river. Some of these, she believed, would level with her. They were people who shared her concern about water quality because they understood how important it was to a healthy fishery. Some she had spoken to about the effects of *Pfiesteria* on fish, and now she said she was pursuing rumors that water quality might be causing problems with human health as well. When they asked her what kind of problems, she steered them ever so lightly in the direction of the symptoms she recognized from her own experience and Glasgow's: open bleeding sores, stomach cramps, respiratory constriction, disorientation, memory loss.

Willy Phillips, a crabber, said he'd never encountered any problems personally, nor had he heard anyone else complain. "Fine," she replied, before asking him where he had his crab pots. As it turned out, they were in an area removed from any identifiable source of pollution.

Jodie Gay, a commercial fisherman, admitted that he had sores on his hands that wouldn't heal, wouldn't respond to antibiotics; and when she asked him where he usually fished, it was in an area close to where she had tracked a toxic outbreak of *Pfiesteria*.

Mrs. Cecil Rhodes said her husband, who both fished and crabbed, had been having health problems for three years: funny infections he couldn't shake, a lethargy that had been diagnosed as chronic fatigue syndrome. And the symptoms seemed to come and go, appearing when he fished in certain areas and when he was cleaning crabs in the summer and early fall, subsiding over the winter. The symptoms were classic, just what Burkholder would expect to see if there was something to this.

These and several other conversations put her in a *This really could be a possibility* frame of mind. The evidence was suggestive and building even before she began to receive unsolicited phone calls and letters that were the result of the press coverage given Howard Glasgow's illness.

A lot of the communications were strange. Apparently the Associated Press had picked up on a story in a local paper headlined "The Pamlico Strain," which described Glasgow's experience and speculated on the human health effects that could be attributed to this organism. Some of the people who called were from Florida or living on the Gulf of Mexico or had vacationed in North Carolina, and they said they were experiencing problems that sounded similar to Howard Glasgow's and wanted to know if it could be related to *Pfiesteria* or if the organism itself might be present in their waters.

To most she said she had no evidence that *Pfiesteria* was active where they were. To the fellow who swore it had to be present in the mountain lakes of Colorado because he was exhibiting symptoms exactly as they had been described, she said, "Even if it sounds the same to you, there is absolutely no way it's the same thing. It's probably some kind of bacteria."

But at the same time she was disabusing people around the country, she was hearing from citizens of North Carolina whose concerns she could not easily dismiss.

She would remember two in particular. The first was a recreational fisherman who said that after a long day of fishing for shrimp in an area with a history of fish kills, he noticed sores on his forearms and hands. Feeling slightly nauseated, disoriented, and too out of it to attempt the drive home, he had retired for the night to a relative's house nearby. Gradually the mental symptoms subsided, he said. As for the sores, they healed more slowly.

Then there was a handwritten letter from an elderly fisherman, who said he had suffered from multiple sclerosis–like symptoms for years before the disease had been ruled out, and he was desperate now because no one could figure out what was wrong with him. He added that he had been crabbing and fishing on the Pamlico since he was a boy.

About this same time she was contacted by a state senator from Craven County in eastern North Carolina, whose office was being inundated with calls from people on the coast with lesions and mysterious illnesses; she asked if Burkholder could recommend a physician.

The only person she could think to refer them to was Don Schmechel, the neurologist who had treated Howard Glasgow. But she knew that his specialty did not prepare him to address the full range of symptoms people were complaining about, so out of frustration at not having any helpful suggestions, she went back to the literature about the effects of toxic marine and freshwater algae on human health.

What she found, after conducting a search of various books, journals, and computer databases, dismayed her. Though certain algal species ranked as producing among the most potent biotoxins known, the information available in this area was partial and piecemeal. In-depth studies had been done on the brevetoxins emitted by the dinoflagellates responsible for red tides, for example, as well as the amnesiac shellfish poisoning caused by shellfish that were contaminated by feeding on toxic algal species. But there remained whole areas about which very little was known. In general, the study of neurological and immunological effects of toxic algal blooms had been less than comprehensive. For instance, there was no national or international surveillance of human illnesses suspected of being caused by toxic algae. It was estimated that the various illnesses algae could cause went unreported by about half because health clinics and physicians simply didn't know enough about what they were dealing with. Burkholder did turn up several studies that showed carcinogenic and immune system suppression properties present in certain dinoflagellate toxins, but there was next to nothing about the effects of low-dose, long-term exposure to these toxins.

It was time to go to the coast and speak to fishermen in person. Her trip was still in the planning stage when Burkholder received a telephone call from Dallas Ormond, a fisherman and crabber on the Pamlico River. Like most men native to the Carolina coast, Ormond talked as if he had a plug of Skoal pocketed in one cheek. Still, the gist of what he was saying was clear enough: there was a fish kill going on, and all his fish were loaded with sores. Could she do something, anything, to bring some relief to the situation?

Recognizing an opportunity to get out on the river, observe a fish kill in progress, and at the same time continue her research on human health effects, she made a date to come down that weekend.

It was a two-and-a-half-hour drive to Bath, a backwater village of colonial vintage whose claim to fame was that Blackbeard had settled there in the 1700s after a career of piracy. She brought Howard Glasgow along to help with some of the sampling she wanted to do, and they met Dallas Ormond at a local marina. He was an unpretentious man with gentlemanly manners, so weather-scuffed that though he was only in his mid-fifties, he could have passed for his father. He was dressed in rubber overalls and white boots, and after loading Burkholder's equipment into his small motorboat, he putted out of the marina, answering her questions as he steered his skiff into the wide expanse of the Pamlico River.

Burkholder began the conversation by asking him if he could tell when a *Pfiesteria* outbreak was imminent. Were there telltale signs in the water? Could he instinctively tell when one was coming on?

He responded that he knew a kill was coming when the water took on "a odor all to itself. Kind of a rotty smell." You could see it too, he said. A "dingy red foam, kinda like shoe polish," would appear on the surface of the river. Fish would be jumping out of the water in front of it as if they were trying to escape a predator, because "anything it catches it kills." Pointing to a Styrofoam float tethered to a crab pot, he added, "Even the barnacles on the underside of them floats."

Spare in detail though his description had been, it was enough to enable Burkholder to picture a night scene, moonlight glistening on the river, perhaps a soft southwesterly wind, the only sound the buzz of a mosquito or the slap of a mullet. Beneath the surface a school of menhaden are moving upriver, filter feeding the way menhaden do, which is by straining tiny plants and animals out of the water with long filaments on their gills. Suddenly a world that is familiar begins to change. Breathing becomes strangely difficult. At first only a few menhaden feel it, and the others notice that some of their own are acting in an unnatural manner, just before they too realize that something is wrong. They don't understand. Some drop toward the bottom, others dart for the surface, but neither place is safe. They are used to fleeing bluefish and stripers to escape being eaten, but this is different. The threat is invisible, and it is all around.

Now some of the small fish are unable to stay upright and are swimming weakly on their side or tilted vertically. Initially they panicked and thrashed, but now they appear to have accepted their fate, without ever having seen their killer. . . .

Ormond fished the way the Indians who once lived in the area used to: with a series of stakes run from just offshore straight out into the river three hundred feet or so, strung with nets to create a series of enclosures that basically trapped fish. He was fishing for menhaden primarily, intending to use them for bait in his crab pots.

"How do you want to do this?" he asked as he killed the motor and drifted toward the stakes. Burkholder told him what she wanted to do, and first she scooped a series of water samples, then she brought out a camera and took pictures, while Ormond and Howard Glasgow— dressed from head to toe in a yellow rubber suit because he was hypersensitized—began to pull in nets, picking up fish and putting them into separate barrels: those with lesions in one barrel, those without in another. When they were finished, they counted and found that of the two hundred fish they had selected randomly, 33 percent had gaping sores on their undersides.

They took a break, watching a flock of cormorants fight over rolling fish carcasses. Then Burkholder brought up the matter of human health effects.

If there was a wind kicking up spray, his eyes and nose would start to tingle and itch, Ormond said. And if he happened to have his hands in the water, the skin around his wrists would burn. When she asked him if he knew of any other people who had exhibited similar symptoms, he referred her to a few old-time fishermen he thought she should speak with, and then he mentioned a situation he'd heard about at Texasgulf.

At the very mention of the phosphate mine, all three of them turned their gaze across the Pamlico to the shoreline, a blue strip separating the river from the sky, except where the silhouette of an enormous industrial complex rose above the horizon in the pall of its own smoke-stacks. Though the view was partially obscured by trees, they could see huge cranes lumbering across bare hills of refuse left over from the phosphate-mining process.

Ormond proceeded to say that a good friend of his who worked for Texasgulf had told him that in 1993, in response to pressure from environmentalists and the state, the company had built an internal pond that was to be used in settling out phosphate dust. According to his friend, employees of Texasgulf soon began to suffer from fainting spells, memory loss, and assorted symptoms that were just what the newspapers were attributing to *Pfiesteria*.

"This is the first I've heard about that," Burkholder commented.

Ormond nodded. "And yer not likely to hear any more. Texasgulf don't want that kind of publicity. According to this here fellow, when they suspected what was causing the trouble, they filled in the pond and told him and everyone who'd gotten sick that they were absolutely not allowed to mention anything to anyone outside the plant about what had happened. It was hushed up good."

"And these people think their problems may have been related to the dinoflagellate they read about?"

"Yep."

"Can you give me the name of your friend, Dallas? So I can talk to him?"

Ormond shook his head. "Nope. I promised I wouldn't. If it got out, he'd be fired. And in this part of the state, Texasgulf is just about the only work you can find that pays. Except crabbing. Sometimes."

For a long time JoAnn Burkholder looked across the river. The cranes looked like dinosaurs moving across a moonscape, and she found herself thinking of them as prehistoric guardians patrolling the perimeter of Texasgulf.

1 3

By the summer of 1994, JoAnn Burkholder's research on this microscopic agent of massive death in fish had been reported on the front pages of major national newspapers and featured in magazines and on television programs. Not only did the adventure of *Pfiesteria*'s life cycle have the ability to inspire deep curiosity; scientists who specialized in the study of diseases and organisms that were emerging as a result of changes in the terrestrial and marine environment saw it as an emblematic phenomenon that dramatically illustrated the limits of our understanding about the microbial world. As a general rule, viruses, bacteria, and other microbial predators tended to evolve toward less virulence, so there would be enough hosts around to eat. The terrifying twist about *Pfiesteria* was that its viciousness did not seem to follow this rule. It killed by unleashing an incredibly destructive toxin, which it continued to release as long as prey was around.

Some scientists went so far as to compare the violence of *Pfiesteria*'s nature, and the devastation it wreaked, to the Ebola virus. What made the "Cell from Hell," as it was described in one publication, potentially more frightening was that it lived not in a tropical rain forest but *here,* in our coastal waters.

Burkholder was also receiving international coverage, particularly in countries that had experienced increasing incidence of "sudden death" fish kills without apparent cause. Clearly this was not just a North Carolina story—the BBC had translated it into thirty-seven languages for worldwide distribution—and as a result of all this attention, she was in demand as a speaker. She was asked to present a paper on her work at a prestigious international conference in Ottawa. She was the featured speaker at various universities across the country. She conducted a workshop in Bodega Bay, California, where she showed scientists how they could recognize *Pfiesteria* in their own waters. For a solid week she was on the road in Maryland, Delaware, and Virginia giving similar workshops and seminars.

Travel defined her life during this period. And adding to her itinerancy were the miles she was logging going to and from her activities as the member of two advisory councils to which she had been appointed by the governor.

The Marine Fisheries Commission was a sixteen-member panel made up of recreational fishermen, commercial fishermen, and seafood dealers, with several at-large seats traditionally occupied by scientists. It was supposed to give the state guidance about how to manage the fisheries in North Carolina, and Burkholder had been asked to fill a vacant seat because of her work in identifying a probable cause of fish kills. Later she was told that she was actually selected because the commission had so many members with vested interests, the governor wanted someone who would be independent, objective, and willing to stand behind her beliefs about what was best for the resource.

Unfamiliar as she was with fishery issues in North Carolina, she had at first been reluctant to accept. The commission also appeared to operate very much like an old-boy system, of which she would be the only woman. But when a top official in the administration said the then-secretary of DEHNR would consider it a personal favor, she did not know how to say no. And as it turned out, her work on the Marine Fisheries Commission gave her an opportunity to pursue her own issues.

Initially she had proceeded cautiously, reading up on the history of the fisheries and absorbing information at the meetings. As the sole female commissioner, she knew that everything she said was going to be scrutinized, and she was heedful of the warning "It's better to be considered a fool than to open your mouth and remove all doubt."

What she quickly learned was that North Carolina's fisheries were in deep trouble. In the glorious 1960s and 1970s, the state's fishing grounds were among the most productive in the world, paradise for both the recreational and the commercial fisherman. But in the early 1980s, many of the stocks became severely stressed or "crashed," and the major reason given by DEHNR was overfishing. There were so many boats and so much good gear being used that the fish didn't have a chance. Just one example was that whereas in the old days fish, detecting vibrations, could avoid the nets dropped by commercial vessels, nowadays sophisticated sonar mechanisms could locate the whereabouts of large schools of fish, and nets could correct for any escape attempts the fish might try. The management question before the commission was how to accommodate the needs of fishermen and at the same time preserve the health of the fisheries in a time of dwindling fish stocks.

Commission meetings were always public affairs, and there were times when a measure was being discussed that was necessary to protect the resource—such as increasing the legal size of the fish that could be caught and kept—but would impose hardship on the fishermen. Wives would plead against the regulation because it would hurt their husbands' ability to provide a livelihood for the family, and JoAnn Burkholder wanted to be anyplace other than where she was. It was heartbreaking for her to have to watch and listen to these people, as they defended a way of life that seemed to be passing, and then turn around and vote for a regulation because it was meritorious for the resource. She believed that commercial fishermen, many of whom came from families that had fished for generations, were part of the heritage of the state and ought to be given special consideration.

But as she became more familiar with the issues facing the fisheries, she realized there were ways she could be of help to them *and* protect the resource. Because overfishing wasn't the only problem facing the fisheries. For years it had been evident that the quality of the environment the fish found themselves in could be having at least as much of an impact as fishing practices.

At subsequent meetings she raised this point, insisting that the commission begin to show more concern about water quality in the fisheries. Estuaries were where almost all the life of the sea began, she reminded everyone. Every important species of commercial fish was dependent on a healthy estuary at some stage of its life cycle, and nutrient-rich water coming from upstream could be devastating to aquatic communities.

It was when she pushed for greater attention to water-quality issues that she learned that the Marine Fisheries Commission had no legal say in those matters. The way the system was set up, that was handled by the Water Quality Section of the Division of Environmental Management. By law, the Marine Fisheries Commission was supposed to concern itself strictly with fish—as if the influence of pollution on habitat were a totally separate matter.

Of course that was silly, and every member of the commission knew it. If fish, crabs, and oysters did not have clean water in which to spawn and survive, everything else that was done to manage the resource was worthless. But traditionally the commission had deferred to the Division of Environmental Management, because aside from the commission's lack of legal authority on most water-quality issues, DEM personnel were supposed to be the experts. When DEM told them that a certain water-quality measure was good, bad, or indifferent where fish were concerned, commissioners would essentially rubber-stamp that opinion, because basically they trusted DEM and there was no one on the commission who knew better.

Now that the commission had someone aboard who was knowledgeable about the science behind water quality—someone who was able to interpret the significance of a given policy recommended by the state and provide scientifically sound advice about its impact on the fisheries—that changed. The commission was able to respond more meaningfully to proposals put forth by DEM, for JoAnn Burkholder, in addition to functioning as a commissioner who discussed and debated issues and voted on resolutions and regulations, was a staff member to whom, whenever the commission considered issues of water quality or anything of that nature, the other commissioners turned for information and advice. Now, if DEM issued new permits for waste discharge into the rivers of North Carolina, or if it released a report on the amount of nutrients it was allowing to be introduced into the estuaries, Burkholder could translate what that meant to marine life downstream.

As a departure from the way things had traditionally been done, this made the Marine Fisheries commissioners feel they could carry out their job more effectively.

Nor was this the only change Burkholder brought to the way water-quality issues were addressed in North Carolina. During public meetings it had become painfully apparent to her that there was a lot of ignorance on the coast. Standing out in relief in her mind was a survey that had been taken as part of the APES process. Asked what they thought was the worst problem related to water pollution, people overwhelmingly replied garbage—floating garbage—which couldn't have been further from the truth. Pesticides, pathogenic bacteria, algae, toxins like dioxin—those were far worse than Styrofoam containers. But garbage was what people cued in on—as if what they couldn't see wouldn't hurt them.

The result was that sincere, well-meaning individuals trying to do the right thing were often rendered ineffective, for they had no idea of the science underlying their concerns. Burkholder would never forget the fellow who stood up at a public hearing on pollution and said, "Them nitrates is burning my fish." He had been right to be concerned about nutrient levels in the estuaries, but his way of expressing his concern and his lack of a factual understanding of what was happening hurt his own cause.

A state official sitting beside her at that meeting had laughed at the man, but Burkholder had sympathized. It wasn't his fault, she thought. It merely highlighted a fact that was true not only in North Carolina but across the country: there was a shameful lack of environmental education among the populace. Decisions were being made that affected people profoundly and that they allowed to happen because they didn't know enough to ask the right questions. Or because they blindly trusted those in charge to act in their best interest. But she'd seen enough to know that all too often the decisions made by the agencies in charge of protecting the environment were compromised by conflicting and incompatible demands. And in many places the decision-making process had become politicized, tilting policies toward what favored powerful special interests.

Knowing there were all sorts of ways that information could be manipulated and decisions described to look as if what was best for the public was being taken care of, when in fact it wasn't; and knowing too that the ability to get away with this depended on an uninformed public, JoAnn Burkholder decided, after that meeting, to try and change things. To do her part in providing environmental education on coastal issues. Which meant taking the time to talk to people around the state as part of an outreach effort, so that they could begin to have a deeper understanding of environmental issues and the problems confronting North Carolina's coastlands.

Voluntarily, she began to address fishermen's associations. She spoke to concerned citizens' groups, to environmental organizations, to high-school teachers. Once, she drove one hundred fifty miles to answer the questions of a six-member garden club.

All of this entailed even more travel, of course, and more time turned over to nonacademic activities. But she considered it a form of service.

Though she didn't like to talk about it much, there was a religious drive behind a lot of what Burkholder did. It was her own brand of religion, and it had nothing to do with church but rather expressed a reverence for nature. She credited her father for its tenets, though they were also summed up nicely in the Book of Revelations: Honor the earth and all its creatures. As it applied to her work as a scientist, and more specifically as a member of the Marine Fisheries Commission, she thought the estuaries and fisheries should be treated with respect. If there was such a thing as a moral absolute, to her way of thinking it was that desecration of the natural environment, and the risk to animals and humans when land and water were fouled by others, was more than wrong. It was a sin.

In large part, the strength of this belief lay behind her acceptance of yet another appointment to yet another advisory council. Soon after his election as governor of North Carolina, James Hunt declared 1994 the Year of the Coast and proceeded to appoint a blue-ribbon panel of prominent citizens to assess the progress the state had made over the previous twenty years in protecting its coastal resources, then make recommendations about how policies should be changed for the better. To many this sounded like a retread of the APES study, and it was. The significant difference was that another administration was taking the initiative.

The Coastal Futures Committee was primarily composed of former politicians, town council members, and city managers, and JoAnn Burkholder was the only scientist invited to serve. Members met once a month in different towns around the state, and to more adequately address the different issues, they formed two subcommittees—one dealing with land use, the other with habitat and water quality—which would meet privately in between the monthly meetings. Because of her technical expertise, Burkholder was assigned to the water-quality subcommittee. Moreover, as the only member with a scientific background, she was asked to tear into current standards and regulations and find out, from the standpoint of fishery resource and water-quality protection, what the state had done, what it was doing, and whether it was achieving its objectives.

What she found, after a thorough reading of state reports, an inquiry into what other states were doing, and assorted research endeavors, was so depressing that, in her view, it justified a sweeping review of all estuarine management policies.

Although North Carolina liked to promote itself as environmentally progressive, Burkholder, going beyond the hype to the facts, discovered a very different story. Few land-use plans considered water quality. Current methods of treating human and agricultural waste had proved inadequate to maintain acceptable levels of water quality. There were serious sediment-loading problems from urban development. Overall, the quality of North Carolina's waters was declining to the point where the water often did not support the animals that depended on it. Data from DEM itself showed that only 18 percent of the 9,300 miles of streams draining into the largest sounds had water quality that fully supported uses officially designated for them. A single statistical comparison said it most dramatically: the state ranked ninth in the nation in toxic discharges, and forty-seventh in spending for environmental protection.

The deeper she dug, the worse it became. She learned that while a lot of rules and regulations on the books sounded good, much of it was propaganda. As just one example, the state proudly pointed to its "designation program" as an indication that it was a national leader in taking water-quality initiatives, so that if a body of water was designated as, say, "nutrient sensitive," nutrient reduction targets were set and special measures taken to bring levels down.

But as near as she could tell, what happened was that someone from the state would walk up to a river and say, "Yup, algal blooms. Must be full of nutrients," and declare it "nutrient sensitive." The state would then make a big deal about how it was going to enact controls to reduce nutrient inputs, leaving the impression that the problem was being resolved. But in fact little changed, because a vital step in the process was missing: enforcement. When point-source dischargers (municipal and industrial) to a nutrient-sensitive waterway were discovered in violation, there was no provision for collecting penalties in a rigorous way, so the fining system was a travesty. As for non-point sources (agriculture), which contributed about 70 to 75 percent of the total pollutant loadings on an annual basis, the widespread approach was voluntary compliance.

The way the Coastal Futures Committee was set up, the in-between meetings of the Water Quality Subcommittee were attended not only by the members but also by representatives of the Water Quality Section of the Division of Environmental Management. The idea was that as relevant issues were discussed, DEM officials would provide staff support, answering questions and assisting the committee as it went about formulating the recommendations that would eventually be presented to the governor in the form of a final report. It was an approach that was sound in principle, but more often than not, the meetings became combative affairs pitting officials from the Water Quality Section against Dr. JoAnn Burkholder, as the only member of the committee with technical expertise.

As Burkholder would characterize her actions in these meetings, when she took the lead in asking tough questions about what was and wasn't being done by DEM to protect water quality, she was only doing what she had been tasked to do. And she says she did her best to choose her words cautiously, pointing no fingers, naming no names. According to her, she refrained from outright criticism of current environmental policies. If there was an apparent shortcoming, she would try to phrase her observation politely, along the lines of: *These are the limitations I've noticed, and we need to begin to address them.*

The tension that resulted, she would say, came about because DEM officials were not accustomed to critical scrutiny. From the start, she believed, they construed the entire process as a threat to their authority, because implicit in the very fact that these discussions were being held was the notion that DEM had not been doing its job.

In defense of this interpretation she would cite specific exchanges that took place in these meetings. When she would ask officials what planning had been done that anticipated the strain on North Carolina's sewage treatment facilities, given North Carolina's massive population growth and the fact that many of the rivers into which treated wastewater was dumped had already reached their assimilative capacity for organic waste; when she would ask what strategies were being employed to rehabilitate areas that had been closed to shellfishing because of previous water-quality degradation; when she asked technical questions about how the division arrived at the levels of nitrogen it was allowing to be discharged into the estuaries when twenty-five years earlier scientists had told the state that nitrogen stimulated phytoplankton growth and tighter standards needed to be established— when she asked these and other probing questions, what she usually got back, in addition to icy stares, were excuses: "Do you have any idea how much money that would cost? North Carolina is not ready to spend that kind of money"; challenges: "The verdict is still out on nutrients. The reductions you are calling for are greater than are needed"; and rhetoric: "We have excellent water quality in our estuaries, and some of the best programs in the country."

Moreover, she would point to DEM's obstructive, as opposed to constructive, participation in the cowriting of a draft report that was supposed to contain proposals for protecting the estuaries and improving water quality. After working for months, the subcommittee would submit to DEM for comment recommendations they wanted to make to the full committee, and what they got back was often unrecognizable. Entire sections would be removed. Recommendations for nutrient standards were gone. An upbeat and self-congratulatory publicity blurb was left, along the lines of "DEM should continue to do what it has done so well."

"What is this?" Burkholder demanded of DEM officials at a subsequent meeting. "If DEM is doing such a wonderful job, let's just all go home. Why are we bothering? These aren't recommendations of what should be done to improve habitats. This is a pat on the back."

One adulteration that especially galled her was what DEM did to her recommendation that effective ways needed to be found to include new scientific findings in the management process. This was important, she felt, because it was common knowledge that there was a lag between what was known in the scientific community and the information regulators used to set policy. But in the DEM-revised version, the recommendation read that since scientists often couldn't agree, it should be left to DEM to decide if and when "science" should be involved in the management process.

This was not how Burkholder thought the state agency in charge of water quality should behave. She thought DEM should use the ideas proposed by the Coastal Futures Committee as leverage in its fight for protecting the resource. That DEM, instead, appeared to resent the recommendations as unwanted interference in the way it did business suggested to her that it liked things just the way they were. And that, she suspected, was because there were so many powerful political constituencies whose interests conflicted with what it would take to improve water quality in North Carolina that DEM, rather than fighting for the resource, was trying to keep everyone happy. Between the lines, that read: The agency that is supposed to be controlling these interests is, in fact, captured by them.

Once she had come to this conclusion, Burkholder grew more vocal about what she thought was going on. She began to put DEM on the firing line. *How did you arrive at your standards? What are you doing to ensure that they are being met? What are you doing when they are not?*

Focused as she was on standing up for what she believed, Burkholder was not oblivious to the ways in which her candor was being received by those around her. Environmental advocates, looking for scientists to validate their particular agendas, embraced her. She was perceived as a champion by many of the coastal citizens, who admired her because she was straight-talking and uncompromising. She was a hot media item because she spoke her mind and did not shy from controversy. But at the same time she could see she was winning public support, she knew she was embarrassing and offending a lot of public officials and state bureaucrats.

To say there was an organized counteroffensive by DEM officials would probably be too extreme. To say, however, that the formidable head of the Water Quality Section, Steve Tedder, saw this as an encore of agency bashing that went back through the Marine Fisheries Commission to the APES study, and took Burkholder's comments personally and was not going to sit still for them, would be an understatement. In the succeeding months, claiming she was making assertions that were not supported by her data, he forcefully waged a campaign to destroy her credibility. He put the word out that she was an alarmist. That she had never been published in reputable peer-reviewed journals. That her scientific integrity was open to question. That she was on a crusade to promote herself in order to raise money for her research. That she had even gone so far as to fabricate her own experience with *Pfiesteria.*

In the arena of political infighting, JoAnn Burkholder was a novice. When she learned about the disparaging remarks that were being made, she was taken by surprise. It seemed never to have occurred to her that officials, smarting at her prosecutorial brazenness, would respond by attacking her and attempting to smear her professional reputation.

The slander that hurt most was the accusation that she was on a self-promotional crusade. Unless she was asked about it, she had deliberately avoided discussing *Pfiesteria* at all her meetings, just so she would not be perceived as using her positions on the committee and the commission as a platform. And when it was brought up, she would always say *Pfiesteria* was just one piece of what was wrong in the estuary. "A harbinger of bad news," was the way she described it.

There were friends and watchers, inclined to be sympathetic, who tried to give her advice. Scientists tend to deal with facts and the search for truth and the way things should be, but that's one of the few places that happens, they reminded her. Politics, on the other hand, is the art of trying to pull people together to do the best they can. Your problem is that you believe you know what's right, and you expect people to go along with you. Would that the world worked this way.

But by this time Burkholder was frustrated and caught up in the conflict, which worked to her disadvantage because it accentuated some of her weak points. She had a way of speaking with such certainty that she could sound absolute in her opinions. When she clashed with others, so intensely would she defend her position that it sometimes seemed as though she was seeing those who didn't agree with her as her enemies.

The truth was that this dispute did sit at the emotional center of JoAnn Burkholder's life during this period. Those close to her encouraged her to take a vacation, and it was a tip she took, packing a change of clothes and heading down to Wilmington, where Mike Mallin held a teaching and research post. Although they did not spend a lot of time together—weekends when she did not have another commitment, an occasional trip to the mountains or to historical sites in Virginia— they were a couple who talked almost nightly on the phone, and there were more than enough good times to sustain a relationship. His support was important to her. It had been her observation that it was difficult for men generally to listen to women who had something on their mind. Men liked to do most of the talking, or else they would respond to a woman's problems with suggestions as to how to make things right. They didn't realize that listening was often the best help they could give. Mike was good that way, and a good listener was what she needed.

～ ～ ～

While all this was going on, Burkholder put in an appearance in the late spring of 1994 at a public meeting in Jacksonville, North Carolina, which would become pivotal to her future. Local fishermen were up in arms over the amount of sewage being dumped in the New River estuary, which was a fragile and important fishing ground. Of the forty-two treatment plants licensed to dump treated wastewater in the river, all but seven were malfunctioning. Burkholder had been asked to come and speak because in the course of her research she had found *Pfiesteria* active in significantly higher abundance at sewage outfall sites on the New River than at control sites away from wastewater sources.

It was not an event she'd wanted to attend—by now she was feeling overscheduled—and she had tried her best to get out of it. But the old fisherman who had once said "Them nitrates is burning my fish" was the organizer, and he kept calling and saying what an honor it would be to have her there.

Expecting it to be low key, attended by a dozen or so concerned citizens, she arrived about five minutes before the program was sup-posed to start. And as she turned into the parking lot at Jacksonville High School, she slowed to a stop, amazed to find it packed with cars. Not only that, but minivans from several TV stations were parked out front.

As she walked up to the building, she was reassessing the old fish-erman's ability to organize, when he rushed up to her. "God, I'm glad you're here," he said.

A tad bewildered, she asked, "What's going on?"

Instead of answering, he handed her a piece of paper. She read it as she walked into the auditorium and up the stairs to the stage. It was a one-page news release that had been distributed to every newspaper and radio and television station in the state just three hours earlier by a county commissioner named Sam Hewitt. He called for the immediate closure of the New River to swimming, boating, fishing—to all human practices—because of Dr. JoAnn Burkholder's research on the killer dinoflagellate and her hypothesis that it could endanger human health.

She looked out at the audience as she took her seat at a table on the stage, and she realized everyone was staring at her. Among the faces she recognized were fishermen, environmentalists, public health officials—camps of people with a stake in keeping the estuary open—and their expressions were stony.

At that moment the TV cameras switched on, floodlighting the stage and letting her know this was not just a meeting of concerned citizens; it was a media event.

Several other people spoke first. Sewage was the announced topic, and local officials complained about a loophole in the law that allowed sewage plants to continue to operate even though they were in viola-tion of existing regulations—if they promised to upgrade the facilities within a specified time period. When it was the turn of the state officials from the regional DEM office to speak, they tried to say, lamely, that like it or not, this was the law, and they were doing their best to stay on top of things.

During a hot debate over the adequacy of what was being done, Burkholder made an effort to look as though she were paying atten-tion, but in fact she was fighting to control her rage. She wanted to strangle this county commissioner, whoever he was. How dare he suggest, without consulting her first, that the entire estuary be closed to human activities because of her research? Given the trouble she was having already with DEM officials who accused her of making unfounded statements, this was the last thing she needed.

When at last it was her turn to speak, she leaned toward the microphone. But the silence in the auditorium was so absolute she could have whispered and been heard in the back of the room.

"There is no point in repeating what some of my concerned colleagues have said," she began. "Along those lines I will only add that it is obvious to scientists that you have a sewage signature in this estuary. The New River is loaded with sewage nutrients, and the fact that the state has designated it 'nutrient sensitive' has clearly done very little to change that status."

Pausing, she looked over the crowd. She knew she was in for some heavy questioning, so rather than try and guess what people wanted to hear her talk about, she decided to let them ask for themselves.

"Perhaps it's best at this time for me to try and answer any questions you have."

Hands shot into the air as though spring-loaded. She would have bet that it didn't matter which person she called on: the question would have been the same.

"Would you please respond to the news release?"

While awaiting her turn, she had wondered if she would be slapped with a lawsuit if she honestly spoke her mind. Now she no longer cared.

"I'd be glad to. I don't know this Mr. Hewitt, but I take it he's not here tonight." A glance over the audience confirmed that the commissioner had not shown up.

"Nor do I know where this individual obtained the information contained in this so-called news release. But I can tell you this: It did not come from me. I don't know why he's doing this. More to the point, it is inexcusable for a public official to take scientific research out of context and use it to cause widespread panic.

"There is absolutely no reason to shut the New River to human activities at this time because of my research. Yes, *Pfiesteria* is a pathogen in the New River. Studies have shown it is stimulated by nutrient enrichment, and it likes sewage outfall sites on the New River. But there are other pathogens in the river. Bacteria. Pesticides. Heavy metals. This is a cause for concern but not for alarm. Calling for the entire closure of the estuary is premature and ludicrous."

At those words, as though air under pressure had been released, there was a collective sigh from the audience.

"Now, as far as human health effects are concerned, I can only tell you what we know. We know there have been harmful effects to people in a closed laboratory setting. But we have no confirmed information yet that it is causing health problems on the estuary. At this point we, as a state, need to respond in a proactive way. We need to go after this organism. We need to understand it better so we can work to discourage its growth. We need to get to know the enemy. We know what makes *Pfiesteria* happy, and that's nutrient-enriched waters. Now we need to begin to take steps to get rid of it if we can, or control it if we cannot. And anyone who goes any further at this time, anyone who uses the research I have done to scare people, must have an ulterior motive. Whatever that is, ladies and gentlemen, it certainly is not that he is interested in protecting you."

A few more questions were asked, but the concern on everyone's mind had been addressed, and officially the meeting didn't last much longer.

For Burkholder, however, the evening ran late. Fishermen and their wives came up and hugged her, thanking her for what she'd done. They had been worried that because of her, they were not going to be allowed to go out tomorrow and do what they'd been doing all their lives. Relieved county health officials wanted to shake her hand, and the TV crews all requested interviews.

The next morning, Burkholder received two unexpected phone calls in her motel room. First, a lobbyist for the North Carolina Fishermen's Association wanted her to know that when he'd learned about the press release he was afraid she would use it to get leverage for her research on *Pfiesteria*. He congratulated her for doing the right thing and pledged his support for her effort to learn more about the effects of this organism on both fish and people.

The second call was from the assistant secretary of the North Carolina Department of Health, Dr. Ronald Levine, who said he had been informed of what she'd done in Jacksonville and thanked her for behaving responsibly in the face of what could have been a public health panic. He assured her that the state intended to be proactive with regard to the potential health effects of *Pfiesteria,* and toward that end he invited her to meet with him when she returned to Raleigh, so they could discuss means of helping her.

Had she received that call a year earlier, Burkholder would have been thrilled. But because the Department of Health, like the Division of Environmental Management, was part of the North Carolina Department of Environment, Health and Natural Resources (DEHNR), and because she knew that DEM officials had been bad-mouthing her and her research, she was wary of everyone associated with that agency.

Cautiously optimistic would best describe her mood as she took the elevator to the fourteenth floor of the Archdale Building, one of three "skyscrapers" in the state's capital, to convene with Dr. Levine the following week. Although she had never met Levine before, she knew he was respected by the people who worked with him. The fact that he was the only assistant secretary who had remained through crossover administrations—Republican and Democrat—suggested he was nonpolitical.

The meeting was held in a conference room and was attended by staff members from the health department and a variety of state officials representing the governor's office. A middle-aged man with graying hair, knowing blue eyes set off by spectacles, and a bedside manner that befitted his profession, Dr. Levine started the meeting off by recounting his personal bout with a toxic dinoflagellate. While attending a health conference in Alexandria, Virginia, he'd fallen ill from ciguatera poisoning after a dinner of red snapper, and it was six months before his symptoms subsided. His own experience, as well as his concern about the public health implications of this newly discovered toxic dinoflagellate, piqued his interest in what Dr. Burkholder had to say, and he asked for an overview of the *Pfiesteria* problem.

She replied that she had been working for more than three years to understand the distribution, ecology, and fish-killing activity of this dinoflagellate; and based on her data, the organism was implicated as the causative agent of 30 to 50 percent or more of the major annual fish kills in North Carolina's estuaries during 1991–93. After describing its horrendous effects on fish, she said that through hard experience, both she and her research assistant had learned this dino could also have serious effects on human health, ranging from weakened resistance to disease (suggesting immune system suppression) to Alzheimer's-like short-term memory loss, with symptoms lasting from weeks to months. She said her assistant's case had been carefully described by neurologists at Duke Medical Center, who had acquired the only in-depth medical knowledge of the toxin's effects on humans.

Over the previous months, she continued, she had undertaken a personal investigation into whether or not there were similar human health effects in the field. While at this stage the information was strictly anecdotal, some of the things she had seen and heard about were worrisome. She knew of more than twenty fishermen who had described sores of suspicious origin, neurological problems for which their doctors had ruled out Alzheimer's and had no other explanation, and other symptoms that suggested linkage to the dinoflagellate.

Based on what she knew, Burkholder strongly encouraged the Department of Health to initiate an effort to evaluate the potential for acute and chronic adverse impacts of this dinoflagellate on fishermen, boaters, water-quality testers, and other people who frequented estuarine areas where it was known to be highly active.

It was an informational meeting—Burkholder did most of the talking—and she had entered it without high expectations of a productive outcome. But by the time she had answered the last question and was riding the elevator to the ground floor, she allowed herself to believe that at last the importance of her work had been recognized within North Carolina. After listening to her carefully, Dr. Levine had said it certainly appeared to him that research into this toxic pathogen was in critical need of funding support. Seeming to be enthusiastically committed to delving into the human health aspects of this organism, Levine had told her that he had some $2 million of surplus funds that would shortly revert to the General Assembly unless the Department of Health spent it, and he wanted to offer her as much as she needed to "really do the research right."

She spent the next week putting together a short preproposal for Dr. Levine in which she itemized a budget that called for about $400,000 to be devoted to the tasks of getting a Level 3 biohazard facility up and running so they could safely mass-culture the organism and obtain sufficient culture for toxin analysis and characterization; mapping the estuaries for hot spots of toxic activity by the organism so they could identify areas of highest potential risk to fish and humans; determining more about the dinoflagellate's effects on finfish and shellfish health, and safety for human consumption; and examining the stimulatory effects of nutrient enrichment on toxic outbreaks of *Pfiesteria*. In addition, she asked for $180,000, the estimated cost of a top-notch in-depth epidemiological study by Dr. Schmechel of Duke Medical Center that would begin as soon as the toxin was identified.

Meeting in a room downstairs from her university office, Dr. Burkholder gave Dr. Levine a copy of the proposal.

"This looks good," she heard him say. "I have a state legislator who is willing to sponsor the request. She will take it to the governor for his sanction, and I'll let you know what happens. It should be no problem."

"How long will that be?" she asked.

"A couple of days at the most," he replied.

Things were happening fast. So fast Burkholder almost couldn't believe the turnaround in her fortune. *Maybe good things do come to those who are patient,* she found herself thinking. And it was in the spirit of cooperation and collegiality that, when Dr. Levine asked her for the names of some of the fishermen she had interviewed who had voiced neurological complaints they believed were associated with the dinoflagellate, she supplied him with the names of four individuals, getting their permission first.

A week went by, and there was no word from Dr. Levine. She did receive calls from two of the fishermen whose names she had given the health department, however.

"Who was that clown you had call me?" the first one snorted.

"What do you mean?" she asked.

"The guy from the health department."

"What happened?"

"I told him some of the problems I was having, and you know what he says? 'Well, you drink, don't you? How can you tell the effects of drinking from anything else?' "

The other fisherman said essentially the same thing; he felt ridiculed and insulted by his caller.

Burkholder was incensed. It was the very reason she had been reluctant to hand over their names. Special consideration needed to be taken when talking to fishermen. They were not inclined to talk about their personal health problems with strangers, especially if they thought what they said might be used to hurt their livelihood. It had taken her a long time to gain their confidence, which was why it was so infuriating to hear one say, "I trusted you, JoAnn, and you send me this? The hell with it. I ain't talking anymore. And I'm gonna tell other fishermen not to bother." Not only was she losing the rapport she had worked hard to gain; the pool for a future epidemiological study was being poisoned.

She was beginning to have a bad feeling about Dr. Levine and his interest in her research. Deciding not to wait any longer, she called him.

Dr. Levine's secretary said he was out of the office, so Burkholder left a message: she needed to know what was going on.

Another two days passed before she received his call. Their conversation lasted approximately a half hour, and her mind reeled as she scribbled notes on what she was hearing. According to Levine, he had checked with officials at DEM, and they had told him that she had such "serious credibility problems" as a scientist that she should not be given the responsibility of conducting research on *Pfiesteria*. On a notepad she wrote: "It is the general feeling at DEHNR that anybody but you should do this research, because you would be considered so biased that your ability to report objective findings would be regarded as questionable." Dr. Levine went on to say that the fishermen who were contacted by his staff by telephone showed no pattern in their symptomology, and that therefore he considered her so-called anecdotal evidence a "joke" without substance. Finally he said that he intended to retain the funds he had requested in his coffers for "other uses," since there was no basis for any further consideration of the entire issue of the potential for this pathogen to cause human health effects in North Carolina's estuaries.

When it was over, she went to her computer. While everything was fresh in her mind, she wanted to get down on paper what Dr. Levine had just said to her. *Stick to the facts,* she told herself, because she found herself wanting to try to recapture the caustic and demeaning tone of the conversation, and she was writing this letter "for the record."

Missing too from what she typed were her suspicions about what lay behind what had just happened. She was convinced that the money Levine had dangled in front of her and jerked away when she reached for it was a way of sending her a message: Back off water-quality issues. Be quiet about *Pfiesteria.* The way she read the incident, Dr. Levine had never intended for her to receive any funds for research. As she saw it, the interest he had expressed had been the culmination of a campaign on the part of DEHNR to show that she was undeserving of trust or belief, and in that way to silence her.

PART FOUR

"Hey, there's something out there.

There has to be."

Part of what had initially drawn JoAnn Burkholder to a career in science was that she felt she did not understand or get along very well with people. Growing up, she'd had little in common with the neighbor kids, and her older sister treated her as a tagalong to be pushed out of playhouses and locked out of cars. As a result, at an age when most children were striving to develop an identity by asserting themselves or establishing their difference, JoAnn was the child who tried to disappear. Her closet was her reading room, the floor of her bedroom a runway she would slide across to a place under her bed that was her haven. Then, in the seventh grade, when the boy who could have won the Mr. Popularity contest developed a crush on her, she had been ostracized by jealous female classmates who spread nasty rumors about her morals, and that was the first time she remembers thinking, *Expect the worst about people, and you won't be disappointed.*

This was why, as a child, she had gravitated toward the animal world. The books that were important to her had titles like *Old Mister Buzzard, Peter Rabbit,* and *Bowser the Hound.* She wanted a dog of her own but her mother wouldn't allow it, so she "adopted" a Pekingese named FeFe, which belonged to an elderly woman who lived down the street. No bigger than a cat, missing most of her teeth, and with a mashed-in face that looked as if it had been battered, not genetically determined, FeFe was a testament to the adage that what counted was the size of the fight in the dog, not vice versa. It was a quality of character JoAnn could relate to.

Even JoAnn's relationship with her mother was sometimes uneasy. Half-English and half-Sicilian, with a classical face and long, curly black hair, Ethyl Galboa was one generation removed from immigrants who had come to this country as stowaways. She had survived a hard-scrabble, Depression upbringing, and at the age of fourteen had dropped out of school and worked in an office until she became a mother and a homemaker. The fact that circumstances had not allowed her to explore a fuller life left her chronically depressed and limited her vision of what was possible for her daughters. Believing science to be a "man's field," she had pushed JoAnn to become an executive secretary; and so often had she complained about how having children had closed off opportunities she might otherwise have pursued, that JoAnn came to view having a family as a trap and vowed, before she could bear any of her own, that she never would.

Her father was the dominant influence in her life. Half Danish and a quarter Irish, with a long face and prominent nose that were attributed to his Cherokee grandmother, he too emerged from poverty-stricken beginnings. After his father deserted the family, his mother had married a man who didn't want to be reminded that he was her second husband, so JoAnn's father was farmed out to a Catholic orphanage when he was just four. As well as growing up with animal books, JoAnn was raised on harrowing tales that could have been written by Charles Dickens: of her father as a young boy thrashed by nuns who didn't know he had a hearing problem, thought he was either retarded or rebellious, and forced him to sleep in a chicken coop; of her father running away on freight trains that took him cross-country through hobo camps during the Depression; of her father hunting rabbit and squirrel with a .22 rifle, knowing that if he missed, there would be no food on the table that night. All by the age of sixteen, when he joined the Civilian Conservation Corps and from there, the army, where he fought as an infantryman in the New Guinea jungle during World War II.

Familiarity with her father's history not only filled JoAnn with sympathy; it helped her deal with his bad temper. Over the most minor things, he was capable of flying into a black rage—and afterward, he would have no memory of what he'd done.

But like the gift of seeing around corners, at a very early age JoAnn developed the ability to recognize that her parents had been shaped by their own childhood experiences, and it helped greatly in understanding them. Besides, offsetting her father's episodic tirades was a closeness born of a common love for the outdoors. When he wasn't working in the foundry, her father wanted to be hunting or fishing, and the change in him when he was tracking a rabbit or reeling in a trout was profound. It was the only time she saw him relax. Young JoAnn realized that in the woods he found himself; *this* was her real father.

In this way JoAnn came to a highly personal understanding of the phrase "at home in the woods." She certainly found it to be the case that plants and animals made better friends than people. And more than anything else, this combination pointed her, as early as the eighth grade, toward her future. Told that if she wanted to make a career out of bird-watching or biology she would need to have something called a Ph.D., she had gone to the library to find out what that meant. And from that day she had known she was going to be a professor of environmental studies.

Just as she was grateful to her father for her connection with nature, now, as she began to feel cornered by North Carolina's health and environment people, she recalled something else her father had given her. Though he wasn't one to pass out advice in the form of long-winded lectures, there were times, driving to and from the woods, when he would relate a story in which a lesson shone through. He'd said once that in the orphanage, a group of older boys had ganged up on him for no good reason and given him a whipping. Even though he was smaller and outnumbered, he had fought like hell. When she asked him why he had bothered to fight back, knowing he was going to lose, he replied, "Because I wanted them to think twice before they came after me again."

Up until this time, JoAnn Burkholder's familiarity with the way politics was played out in a scientific setting came from the behavior of fellow grad students in the lab where she received her doctorate. But that had been petty stuff: playing up to the major professor, ratting on fellow students who said anything derogatory about the way the lab was run. The State of North Carolina's actions in this instance suggested to her that science was being politicized on a higher level, and in her mind it went straight to the very issue that she had argued needed to be addressed during the Coastal Futures process: how government handled scientific findings that bore directly on established policies and practices.

She had a bad feeling that something sinister was going on. As she put it together, because she had taken a leading role in the public debate over the quality of water in North Carolina and made little effort to disguise her low opinion of the job the state environmental management agency was doing, officials within DEHNR were countering with a calculated effort to bury the issues she raised and quiet her. And apparently they had decided that the most effective way to do both would be to discredit her by accusations of bias and irresponsibility. She believed they had been waiting for the right opportunity, and that it had come along when she moved beyond the notion that *Pfiesteria* was a highly visible symptom of water-quality problems, to the possibility that it was a potent pathogen whose impact extended beyond North Carolina's fisheries, threatening fishermen and other coastal citizens. By pretending to be serious about investigating the nature and extent of the threat, then assailing her professionally, raising questions about her scientific integrity, officials had engaged in the ultimate put-down. When she looked at the way they had gone about it, it followed as relentlessly as a logical conclusion that Dr. Levine's actions were part of a larger effort by high officials within DEHNR to put a halt to her career.

It was no exaggeration to say that at this point JoAnn Burkholder perceived herself to be fighting for her professional life. Or to say that just as her father had been a street fighter willing to take on a challenge no matter how badly the odds were stacked against him, so she drew a certain amount of strength from thinking that she was too. It was an advantage she felt she had over state officials. They were bureaucrats; they weren't street fighters.

Of foremost concern was the slander issue. She was afraid that if she did not stop it right away, it would spread and poison her ability to raise research money elsewhere. So the very next thing she did was pick up the phone and call the deputy secretary of the North Carolina

Department of Environment, Health and Natural Resources. Steven Levitas was a political appointee of the current administration. Previously an attorney with the Environmental Defense Fund, he had high expectations from environmentalists. But to Burkholder he had proved a disappointment, because the department had seemed to pursue business as usual.

Personally, she had found Levitas to be insufferably arrogant. On several occasions, when she told him she was afraid citizens on the estuaries were being hurt by *Pfiesteria* and he needed to take some initiatives, he had all but laughed in her face. After one Coastal Futures meeting that he had attended she had tried again, saying, "I want you to take this seriously. I won't keep taking no for an answer." To which he had retorted, "Well, as I understand it, short-term memory loss is one of the symptoms. Are you even going to remember this conversation tomorrow?"

Now, when Burkholder identified herself and asked to speak to Levitas, his secretary said he was in a meeting and could not be interrupted. Sensing her call was being screened, Burkholder snapped, "You tell him he's got ten minutes to get to a telephone. If I don't hear from him in ten minutes, tell him I'm going to the press and will blow the scheming that is going on in his agency nationwide."

The phone rang in eight minutes.

"What's the trouble?" Levitas said, innocently.

But it was inconceivable that midlevel officials in his agency could be waging a campaign against her without the knowledge and sanction of people higher up. When she called him on it, he admitted that he had seen E-mail going back and forth between the health department and DEM in which her name was mentioned, but said he had not paid much attention to what was crossing the screen.

I'll bet, she thought.

"Plead ignorance if you want, I don't care. But you're the deputy director of this agency, and you ought to know what's going on."

"Really, I don't," he protested.

"Slander is coming out of your agency about my reputation and scientific credibility, and I want it to stop or I'll take legal action."

"How am I supposed to—"

"I want an emergency meeting of all interested parties. If I don't get it I'm going to the press. And if for a minute you don't think I'm going to do it, *watch* me!"

She got her meeting the next afternoon, in the Archdale Building in downtown Raleigh, in the same room where this whole debacle had started. But she received very little satisfaction.

Levitas was a no-show. In attendance, however, were various officials within DEHNR, along with Mrs. Marion Smith, a former legislative aide who represented the governor of North Carolina, James R. Hunt, Jr., on issues facing the eastern part of the state. And of course, Dr. Ronald Levine was there.

The slander issue being at the top of her list, Burkholder went through what had happened, repeated what she understood was being said about her credibility, and reviewed her qualifications. Specifically, in response to the charge being circulated that her research had not been published in peer-reviewed journals, and to the fact that a "letter" to the internationally renowned scientific journal *Nature* had been described by officials in DEM as a mere "correspondence," she offered enlightenment.

" 'Letters' to *Nature* are short scientific articles concerning new findings of major importance. They are considered sufficiently important that scientists regard them as significant achievements. *Nature* has a 99 percent rejection rate of its 'Letters,' because the publications are critically peer-reviewed."

A certain amount of demurral occurred at that point. One of the officials suggested that she must be exaggerating, no one in DEM would have said such things, but Marion Smith spoke up and confirmed that she had heard those very accusations made about Dr. Burkholder in public forums on the coast by DEM people.

All eyes then turned toward Dr. Levine, who Burkholder said had repeated those accusations to her in their phone conversation and who now, to her, seemed to assume the party line with a patronizing twist.

"Dr. Burkholder must have misinterpreted what I was saying. I was just trying to warn her that there were people here who didn't really think very highly of her and there was a credibility gap in her research that she should be aware of and try to overcome. I was trying to give her advice about what was in her best interests."

Glaring across the table at Levine, it was difficult for Burkholder to restrain herself. Their phone conversation had been so insulting to her that afterward, whenever his name passed her lips, the expression on her face would curdle.

So you're saying I'm making the whole thing up? she thought. *What would a professor in my position have to gain from making false charges against a ranking public official?*

After some discussion, she realized that she would not get the promise of a public apology that she wanted. What she got were vague assurances that the sources of this misunderstanding would be given more accurate information about her publishing history.

From there the conversation turned to the matter of funding, and here again Burkholder presented her understanding of what had happened: that after she had expressed her concern to Dr. Levine that an important disease vector for the fisheries had been linked to serious health effects in the laboratory and was worthy of scrutiny in the field, through his direction, using a proposal she had written for funding support for her laboratory, some $600,000 of unspent funds from the Department of Health that would have reverted to the General Assembly were retained specifically to fund her research and an epidemiology study. And that once the funds had been retained, Dr. Levine had reversed his position and decided that even though she was the scientist most experienced throughout the country in working with this pathogen, she should not be given the responsibility of leading the research effort because her ability to report objective findings was regarded as questionable by his agency.

In response, Dr. Levine once again took the position that Burkholder had misunderstood him. He said that he never had any intention of handing over to her more than a half-million dollars. What he'd had in mind all along, he said, was that the moneys would be administered through a third party, who would distribute them to the most qualified researchers through the RFP process.

Burkholder cast a smoldering look his way. The RFP was the standard protocol for distributing grant money within academia. An agency that had a scientific question it wanted answered would issue a Request For Proposal describing its need, interested investigators would submit proposals, which were sent out for anonymous peer review, and the results were presented to a panel of advisers, who looked over the reviews and ranked them. Funding was awarded on the basis of rank order. Certainly Burkholder was familiar with the process, indeed it was the method through which she had received most of her grant moneys, and in principle she had no objections. She thought this case was different, however. Levine was changing the "rules" in the middle of the game. Where at first he'd led her to believe that if the Department of Health succeeded in locating the money, a significant amount would be earmarked for support of her work, now he was born again on the issue of competitive bidding and wanted the money to be part of a general call for anyone who wished to enter this research area. She had never thought that he would simply hand the money over to her, as if he carried it in cash in his wallet. She had thought it would be contracted. State agencies that were not equipped to do specialized research frequently contracted directly with university scientists for specific work. It was done all the time, and it was what Dr. Levine had implied would happen here.

Measuring the anger in her voice, she said, "How will I know, based on what's happened to me here so far, that this will be a fairly conducted process?"

"Well, all I can do is try to find someone you trust to oversee the RFP, Dr. Burkholder," Levine said. "Because it's obvious you don't have much trust for most of us here."

When she didn't reply, he added, "You're not afraid of an RFP, are you? I mean, since you are the premier researcher on this, I would assume that you would compete fairly well, don't you think?"

She didn't know what to say. It wasn't the ability to compete that concerned her. Normally she would have felt perfectly comfortable competing for funding in a general call for proposals, since her track record at writing successful grants was strong. It was the manner in which Dr. Levine and state officials had conducted themselves that convinced her that when it came to her, they were capable of acting in an underhanded manner, and that fairness would be a casualty of any RFP process that was connected to them.

Feeling checkmated, however, she agreed to participate in an RFP process. But she was unhappy—indeed sickened—about it. And during the next several weeks she brooded over the outcome of this meeting, because the more she thought about it, the more she experienced a real sense of betrayal. And yet she felt helpless to do anything but wait and see what happened next and respond to it.

She didn't have to wait long. In mid-July, she received a polite letter from Dr. Levine, advising her that he was forming a "Scientific Oversight Committee" to work out a comprehensive epidemiological approach to studying the potential health effects of *Pfiesteria*.

Funny, she thought when she'd finished reading his letter. *When we talked on the phone, he told me he didn't think an epidemiological study was worth doing. Now he's convening a group to advise him on how to proceed with a study.*

It also struck her as odd that he had apparently chosen not to include Dr. Schmechel, the only man who had clinically treated someone known to have suffered from exposure to *Pfiesteria*'s toxin. Nor had Levine invited her, the person he had acknowledged knew more about the behavior of this organism than anyone else on earth.

She felt extremely uneasy about this meeting, assuming that nobody who attended would understand the issues very well or, if Levine was in charge, care about them.

Several days later, she received a polite letter from Dr. Levine, saying he wanted to be sure and keep her apprised of what was going on and including a copy of the official minutes of the meeting. She read them slowly and carefully, noting how wonderful everything sounded on paper, while thinking as she read, *You no more want to keep me apprised of what's going on than birds fly north for the winter.*

When she finished reading, she decided to raise the stakes.

Several months earlier, she had been notified that the Alumni Association of North Carolina State University had selected her to receive an Outstanding Research Award. In addition to the prestige, a cash allotment of two thousand dollars accompanied the award, both to be presented at a luncheon ceremony. This was welcome news for a change, but it created a scheduling conflict. One of Governor Jim Hunt's aides had contacted her and said that the governor was speaking at a conference in Wilmington dealing with North Carolina's oyster fishery on that day and wanted to meet her. Although Hunt had appointed her to the Marine Fisheries Commission and the Coastal Futures Committee, it had been on the basis of staff recommendations, and they had yet to meet face-to-face.

Trying to squeeze both events in, she drove to Wilmington early that morning and, immediately after she was escorted onstage and shook the governor's hand, jumped in her car and raced back to Raleigh for the awards ceremony. *All for a photo op,* she kept thinking, sure that she would be tagged with a speeding ticket. But now the event in Wilmington took on a different cast, because when she shook the governor's hand, he had leaned toward her and whispered, "If you ever need any help with your research into this fish-killing dinoflagellate, please let me know." And taking her hand in both of his, he added, "I mean it."

Writing the governor a letter was a big step, especially when it was to bring to his attention a problem that existed within a state agency. But it was a measure of her frustration and indignation that she sat down at her computer and did just that.

It was a difficult letter to write, and she made several drafts. The end product was two pages, single spaced, and she wasted no time in getting to the point: "In this letter I am requesting your consideration of intervention in my behalf on an issue in which, I believe, certain NC DEHNR staff have shown unethical conduct. I am uncertain as to whether these staff may be acting in the best interests of the State, and I appeal to you—please help."

She went on to review how the situation she faced had come about and to state her opinion that immediate research was needed to determine the extent to which this "highly toxic new organism" threatened the health of people living on the coast. After pointing out that "My laboratory is one of only two laboratories in the world which has worked on toxic stages of the organism . . . and I am internationally recognized as the foremost scientist with expertise in culturing it and understanding factors which stimulate its toxic outbreaks," she said that it was only logical that she be the leader in the effort, which it was in the state's best interest to accomplish as quickly as possible. "My fear," she concluded, "is that the funding will be spread thinly in a diffused effort among various researchers who would nearly all be inexperienced in working safely with the dinoflagellate; and this course would block indefinitely any chance for me to make real progress in answering questions that our fishermen, tourists and coastal citizens are anxious to know." She ended by saying: "I believe that only your immediate intervention or guidance can now prevent this course from becoming a reality."

She did not trust the mail—she was afraid some staffer would read the letter and it would never get to the governor—so she decided to hand-deliver it. After several calls, she learned when the governor was next scheduled to speak in public, and on that day she put on a blue-flowered skirt, a cream-colored blouse, a navy blazer, and heels, and with the letter in hand, she drove downtown. In the Legislative Building, she wandered up and down the halls until she found the room where the governor was to speak.

She'd come early, wanting to hand him her letter before he stepped up to the podium, because afterward she was afraid he would be surrounded by people. As it turned out, the meeting had begun and someone else was speaking, so to settle her nerves, she took a seat on a bench in the hall and for the umpteenth time proofed her letter. And she came across a sentence she regretted including.

Thinking, *Oh, dear, I can't say that,* she hurried down the hall and out of the building and back to the garage where she'd parked her car. Speeding to her office at the university to make the change and print a revised copy, she then raced downtown again. . . .

From the back of the room she spotted the governor, and she realized she had no choice but to wait until the meeting broke for lunch before she approached him. Working her way toward the front, she took a seat that positioned her strategically close, and then she listened to several speakers talk about health care for the elderly before finally the governor spoke. When he was finished and the applause had ended and the meeting broke up, she walked straight up to him.

"Excuse me, Governor Hunt," she said.

He turned, and it was apparent he didn't recognize her.

"I met you at the oyster summit in Wilmington a few months back," she said, "and you told me if I ever had any problems conducting research on a toxic organism I was working on, to let you know. I've written you a letter, and I wanted to make sure you got it."

When she handed it to him he said, "Thank you, I'll look at it." But he was abrupt, and she noted that rather than sticking the letter in his pocket, he put it on top of a stack of material.

On her way back to her car, all she could think was, *Well, I've probably just slit my throat.* But having proceeded this far, she felt she had no choice but to go the whole way.

~~ ~~ ~~

In her letter to Governor Hunt, Burkholder had tried to make the point that this previously unknown pathogenic dinoflagellate was a "special circumstance," which carried with it a time factor. "This subject is very different from a general sewage problem or general fisheries question in which many people have expertise and suitable facilities, where general (and time-consuming) calls for proposals are standard procedure," she had written. She hadn't wanted to alarm the governor, but she did hope to impress upon him the very real possibility that people on North Carolina's coast were at risk right now, and she had done her best to inform him that if the moneys were distributed as she believed they were intended to be when they reverted to the health department, the support they provided would enable her to get her lab, which was still shut down, up and running—an essential first step in the research process.

A week passed before she heard from the governor, and her heart sank. "I share your concern about the potential effects of the dinoflagellate toxin," he wrote. But he went on to say things like: "It is my understanding that you suggested, and agreed, that these funds would be administered through" a program involving peer review of proposed research projects; Dr. Levine has "assured me that the protocols for the investigation of toxic pathogens are standard and will be fully implemented as warranted"; and "I would appreciate your cooperation with the State in facilitating the funding process so the problem may be addressed in a timely manner."

It was obvious to her that the governor knew only what he'd been told and that Dr. Levine and perhaps other officials in DEHNR had described the situation so as to support their position. She had never suggested that the funds be administered by a third party. Why, if the Department of Health had originally been willing to contract with her directly, would she say, "Let's open this up for competition"?

Her spirits were at an all-time low. She was afraid that writing the governor had done nothing but hurt her and things would only get worse for her with DEHNR. She was not even consoled when the governor's representative for the eastern part of the state tried to be encouraging.

"What on earth are you worried about?" Marion Smith said. "There's nobody who can compete with you. With your expertise, you are clearly advantaged in any competition for the funds. Besides, the RFP is good scientific practice. That's the way we're supposed to do things with tax money. So write your proposal, and let's get on with it."

"You don't know how they can manipulate an RFP," Burkholder replied. "They will craft this whole thing so it will be slanted away from anything that will make real progress. They will fix it so that I will not get this funding."

If Marion Smith's optimism had a tinny ring to Burkholder, Burkholder's concern about the slicing and dicing of the grant money struck Marion Smith as slightly paranoid. Personally, she liked JoAnn Burkholder. Professionally, she admired her achievements, as a woman and a scientist. She felt that unlike a lot of academics, Burkholder had spent time educating herself about issues that lay outside her particular field. About wastewater treatment plants, for example, and the difference between industrial and municipal waste and how they affected water quality, Burkholder was, in her opinion, better informed, outside of a few engineers, than almost anybody in the state. But Marion Smith also recognized that Burkholder was determined to believe that what she suspected was going on with DEHNR had to be true. And at this stage Smith had no reason to think that there would not be a fair hearing, conducted in a timely fashion, and that when the outcome was announced, the bulk of the research money would go to Burkholder's laboratory.

She had yet to realize that if JoAnn Burkholder wasn't sure about something she would usually admit it, but when she felt strongly about an issue she was usually right. And about what was going on behind the scenes at North Carolina's Division of Environmental Management, she was most definitely right.

Although officials in the Department of Environment, Health and Natural Resources were frequently quoted as saying that North Carolina was doing a top-notch job of protecting natural resources and environmentalists who said otherwise were guilty of a sensational distortion of the facts, a compelling body of evidence argued that those officials were deceiving the public. And nowhere was the discrepancy between myth and reality more obvious than within the Water Quality Section of the Division of Environmental Management, which maintained publicly that there was no significant pollution going on in the state's rivers and estuaries, at the same time that almost every reliable indicator said water quality was in serious decline. Put on the spot, DEM officials would point to improvements in certain areas; but it was clear to anyone who looked closely at the record—of shellfish-bed closures, loss of aquatic grasses, increase in dissolved nutrients, high densities of harmful bacteria, toxicity levels in seafood —or spoke with staffers whose loyalty was to the environment, not to higher-ups, that there were major problems with the way DEM went about the business of safeguarding the state's waters.

Across the nation, scientists had reached a unanimous conclusion: the major threat to the health of coastal waters was excessively high concentrations of nutrients. But officials within North Carolina's environmental bureaucracy had been exceptionally slow in acknowledging that link. Their initial reaction had been to deny or dismiss the information when it was presented to them, or to state that they had developed their own opinions on this issue and on what basis they felt action was required, and everything was under control. What it amounted to was that they did not like their authority questioned.

For good reason, according to their critics. The most generous explanation was that the bureaucracy was geared up technically to deal with water-quality issues in a certain way, and there was ownership associated with their way of dealing with the situation. The moderate explanation was that DEM worked the same way as most government agencies—nothing changed until there was a crisis, and steps that brought about incremental improvements allowed them to say they were moving in the right direction. Cynics alluded to more going on —powerful political and economic interests didn't want these issues dealt with because it cost money to clean up pollution, and the Division of Environmental Management, bearing in mind the state's commitment to economic growth, was unwilling to do what was right for the resource if it was politically unpopular.

Probably the clearest area of breakdown within the environmental management system lay in the way science was used. DEM maintained publicly that the most recent scientific information was continuously considered when it came to developing environmental policies and making regulatory decisions. But the experience of a number of scientists who had worked with DEM had been that in fact the managers and regulators within the agency were hostile toward current science —especially science that connected anthropogenic eutrophication (man-made pollution) to ecological stress.

In the scientists' opinion, the people in charge were conservative to a fault. A problem came up, and they waited for it to go away on its own. Only when they absolutely *had* to address it did they; and then one heard statements to the effect that there was not enough scientific evidence to prove irrefutably that the problem was serious. When data were presented to them that established just the opposite, they would assume various defensive positions: There was no basis for scientific consensus. Those particular findings could represent a onetime cycle, not a trend. The projections from our computer models indicate we are on the right track. More studies are needed before we can switch to a new plan of action.

And when they found themselves confronted with scientists who were willing to stand up for the science, agency higher-ups would do their best to marginalize them within both the scientific and the regulatory communities, accusing them of making assumptions outside their area of expertise, or of being disgruntled because they could not "translate pet theories into public policy."

It was a rare scientist who came forward. Most scientists in North Carolina, and in particular those who depended on the state for some part of their funding to continue their research, handed over data and made conclusions about their findings, but their comments usually stayed comfortably within the strict parameters of the scientific study. Seldom did they make a recommendation about what they thought needed to be done.

Behind their reluctance were understandable reasons, survival being foremost. Scientists who got mixed up in the broader implications of policy could conceivably be embroiled in controversy, thus jeopardizing their ability to go back to the funding institutions for more money.

This was what was so unusual about JoAnn Burkholder. She made her work relevant. She was willing to draw policy conclusions from what she was doing and make recommendations. She was looking at the nutrient issue, and she was clear about what the impacts were: the state's waters were polluted to a high degree, and not very much was being done about it. More important, she was willing to take the next step and say, "We need to come up with a systematic way of reducing the nutrient loading in our rivers to protect our estuaries and reduce fish kills."

Rather than embracing this candor, DEM had attempted to use her outspokenness against her. She was overstepping the bounds of her science, they said, revealing her real motivation, which was straight environmental advocacy. Ergo, she was not objective.

There was only one problem with that charge. When Burkholder spoke about policy, she tried to make it clear that she was speaking in her capacity as a member of both the Marine Fisheries Commission and the Coastal Futures Committee, where it was her responsibility to address policy issues. It happened that she was also a scientist whose research informed her opinions—an important distinction.

There were witnesses to this conflict who were inclined to say that DEM's approach to water quality, as well as its attitude toward Burkholder, might have been due to something inherent in public agencies. A certain orientation had emerged over the years, they pointed out. Like any other bureaucracy, DEM had invested in institution building, and with that had come a go-slow approach to problem solving that was a result of political realities, the major one being the state legislature, which represented powerful segments of the state economy and controlled the budgets of departments like DEHNR.

But others denied that was the problem. If anything, they said, it was a mind-set that had been put in place by the leadership of the department and perfected by Steve Tedder, the chief of the Water Quality Section.

Tedder was a career bureaucrat who, without benefit of an advanced degree in the sciences, had been the titular head of Water Quality in the state since 1989. He was a burly man with heavy jowls and wavy brown hair. On a shelf in his office in downtown Raleigh was a bottle labeled "Grouch Pills," which had been given to him as a joke, but the joke that went around was that he actually took the pills, and that they explained the perpetually grumpy expression on his face.

Tedder's management style was very much command and control. He was a man with rigid ideas about how water-quality problems should be handled, and he was unreceptive to people who disagreed with him. It was a widely known fact that once he made up his mind, you weren't going to change it unless the evidence was absolutely overwhelming, and no one could remember the last time that happened.

Environmentalists felt he was too close to business lobbying interests, a suspicion given currency by the fact that his predecessor was currently employed as a lobbyist for the chemical industry, and the person before that represented the North Carolina Homebuilders Association. And it was a fact that at public meetings concerning coastal development, Tedder did seem overly sympathetic to those who declared that more regulations were "environmental luxuries" and could lead to legal haggling because they often conflicted with the rights of private property owners.

Others said no, he was just committed to the policies that were in place, which he genuinely thought were adequate, and if he had a fault, it was that he was overly concerned about the political implications of the remedies to environmental problems.

The one thing about Tedder that everyone agreed upon was that no one upset his sense of order quite like JoAnn Burkholder. As one of his staff people put it, "Burkholder isn't a four-letter word. But whenever Tedder said it, it sounded like one."

An incident that captured the character of their relationship occurred in Southport, a small North Carolina fishing village where Burkholder and several members of the Coastal Futures Committee were supposed to meet with Steve Tedder to discuss his reaction to a series of recommendations for protecting water quality. Burkholder came early to the courthouse where the meeting was to take place, and after a brief exchange with several committee members, she left the room. By the time she returned, Tedder had arrived with his staff, but he too had gone back out of the room, leaving on the table in front of his seat a stack of the recommendations the committee had put forth. And scrawled across the top of every one in black capital letters was the word "BULLSHIT."

While it sometimes appeared that she was doing battle with the entire Department of Environment, Health and Natural Resources, throughout this period Burkholder was in fact having encouraging conversations with some very good DEM personnel who worked in different regional offices: biologists and chemists, primarily; people actually working in the environment, implementing programs, making analyses, and sending reports to the central office in Raleigh. And from them she heard all sorts of horror stories about the way the division was run, which differentiated between what the field people were seeing and what officials in Raleigh were saying.

From people whose job it was to monitor levels of pollution in the state's rivers, she heard about various scams:

How data gathered by field biologists identified certain industries and wastewater treatment plants as the source of pollutants that exceeded state limits, but when it was sent to Raleigh it didn't seem to impress anybody, because nothing was done to change things and no one understood why until it became obvious: the state was getting federal money from EPA to run a surveillance program but not an enforcement program.

How the state had not kept up with available scientific knowledge and technology and was relying on obsolete methods and inadequate indicators for information on which to establish regulations. As a result, the tests they performed and the measurements they took were not sophisticated enough to disclose data on many of the complex and harmful chemicals that industry was discharging into the environment, or the range of human pathogens present in the water. This, in a state that, according to the Environmental Protection Agency, in 1992 had the dubious distinction of placing six companies on the list of the top ten in the country that released the most toxic chemicals into the environment.

How field personnel, after taking physical and chemical measurements in an estuary to determine the nutrient levels that stimulated and supported algal growth characteristic of eutrophic waters, would pass this information up the chain of command with the understanding that it would form the basis for the division's justifying stricter regulations—only to be told their figures were in conflict with economic realities and to come back with numbers more realistically achievable. And if they complained, "This is outrageous: the system responds to an order of magnitude lower than what you're asking for; the technology is available to control this; the problem will only get worse if we don't do something," they were told that regardless of the facts, regardless of their opinion, this was going to be the policy, so get with the program.

People know what's going on, Burkholder was told. *Policy is driven by money and resources and politics, not the needs of the resource. But few of us talk about it because rebels have a short career in this department. It's a good-old-boy system, and there's an unwritten rule that people in the department have to stick together.*

There was a story told by DEM staffers about a meeting between a director from Raleigh and a regional supervisor, who was told, "Remember, you're responsible for making me look good. Make me look good, and I'll take care of you." And he was serious about it. But when word about it got around, it turned into a joke. "Well, I've got to get up this morning and go to work and make the director look good," people would say. "Are we making the director look good today?" became the measure of time well spent. It got to the point where headquarters issued an edict—no more jokes at the director's expense.

A lot of what we do is a front, Burkholder was also told. *We get away with it because there is no accountability, and the people at the top are good at dissembling. At presenting equivocal situations in the best possible light. At presenting the public with a deceptive selection of the facts.*

None of this surprised Burkholder. Not even the admission that DEM would knowingly mislead the public. At a hearing in New Bern in the fall of 1994, at which citizens reacted to a proposal by a small upstream town that wanted to upgrade and expand a sewage treatment facility that had been malfunctioning, she had sat in the audience, listening to angry coastal people oppose the request on the grounds that there should be no more discharges allowed into the Neuse River —which was already designated "nutrient sensitive"—when a state official stood up and tried to reassure everyone that DEM was allowing the plant to open only after it installed improved controls. He went on to say that by subjecting the sewage to a bubbling process that increased oxygen levels, the ammonia, which was a key form of nitrogen in sewage, would be significantly reduced before the wastewater was released into the river. And to show everyone how successful these controls would be, the official displayed a series of colorful acetate charts that illustrated the levels of ammonia before and after treatment.

"Look at these numbers," he said proudly. "Yes, we're expanding the facility to allow for population growth. But the nitrogen levels will be a lot lower than they were. What we're going to do here, in the end, will actually help the Neuse River."

Sitting next to Burkholder was a professor from East Carolina University who was knowledgeable about nutrients. Nudging her, he said, "Do you believe this?"

She shook her head and muttered, "Jesus."

He laughed and said, "I'll flip you for it."

Taking a quarter out of his pocket, he flipped it and she lost.

When the time came for public comment, Burkholder strode up to the microphone and broke the bad news.

"I have a great deal of sympathy for the town officials who want to expand their sewage facility and seem to genuinely want to do the right thing for the estuary by lowering nitrogen input," she said. "And it's my understanding that you have relied on the advice and figures provided by DEM. Is that right?"

The city manager, who was seated onstage, nodded his head.

"That's what I thought," she said. "And this is really sad, because the truth is not being presented to you. You have been shown ammonia data and been told that this treatment will reduce its nitrogen content. But, gentlemen"—and here she turned and looked directly at the state officials—"where are your nitrate data?"

No one moved, no one spoke.

"Now, I'll tell you what these people have done. Yes, ammonia levels will go down when you aerate sewage. But all that happens is the ammonia gets converted into nitrate. I challenge the state to prove that the actual nitrogen loading is going to be decreased, because you're not taking out any nitrogen at all, are you, gentlemen? You are simply changing its form."

When none of the officials responded, Burkholder addressed the audience.

"If you just convert ammonia to nitrate, you don't do anything to remove nitrogen. It may be slightly less toxic to fish, but you're going to get a lot more algae, who will absolutely love you, because they love nitrate. You are being led down a primrose path here by people who are not leveling with you. This will not reduce nutrient levels, and the pollution in the estuary will only get worse."

When she walked from the stage the crowd applauded, while DEM officials shook their heads and made faces, as though she hadn't been telling the truth. But the very next day she received a call from a DEM staffer in a regional office, who said, "JoAnn, I just wanted to call and tell you that was right on. For years that's what some of our people have been doing. I've tried to tell them, but they won't listen, and it's driving us crazy. It's so bad, I've heard that the people at headquarters take bets before these meetings on whether or not the public can be duped again on this nitrogen situation. I know it's hard, but please, keep doing what you're doing."

If nutrient loading challenged the state agencies and the regulatory process in ways it had difficulty dealing with, then it was no wonder that officials saw nothing but trouble coming when JoAnn Burkholder popped up with *Pfiesteria*.

It would be difficult to date the first "official" sighting of *Pfiesteria* in the wild, but a case could be made that James Guthrie, a red-haired technician with the National Marine Fisheries Service in Beaufort, was one of the early ones to see it. Back in the fall of 1984, as part of a NMFS program to obtain information it needed to manage the fishery, he was tagging juvenile menhaden in their nursery grounds in Hancock Creek, a tributary of the Neuse River that spilled into Pamlico Sound, when he noticed something very strange. A large number of the fish he saw were missing a half-moon chunk from their bellies. At first he hadn't paid too much attention, attributing it to bluefish bites; but as he continued to tag, he continued to turn up more and more fish with holes in them, until the numbers were so abundant that he could not ignore the fact that this was different. After looking more closely at the wounds, he discarded the predator theory because there were no signs of a tooth slash. Besides, all the holes were the same shape and every one was in the same place. *Does this signal the outbreak of a new disease of some sort?* he'd wondered. *Could some intestinal parasite, consumed by the menhaden, be eating its way out from the inside?*

When he reported his finding to the biologists back at the Marine Fisheries Service laboratory, they said fish died all the time for unknown reasons. So a few were queerly marked: that was no cause for alarm.

But Guthrie thought it was important. A coastal native, he'd been around the water all his life, and he knew this indicated something was wrong with the fishery. But he couldn't get anyone to take him seriously.

Months passed, and Guthrie could not let go of what he'd seen on Hancock Creek. And he continued to see more "anomalies" in the coastal waters. More fish with half-moon wounds. Blue crabs with mysterious holes in their shells. Even eels were being eaten up, and nothing he knew about attacked eels.

Then came a tagging trip on Stow Creek in New Jersey. Standing in the bow of his boat, scanning the water for signs of menhaden, he smelled something foul in the air. He recognized it as the stink of decaying fish, and it seemed to be getting stronger by the minute, as though carried on a wind. But the air was absolutely still.

He was trying to figure out where the odor was coming from, when he spotted a school of menhaden working their way upriver. Tossing the cast net into the middle of them, he began pulling it in hand over hand and peering into the water, then he saw something that startled him so badly he lost his balance and fell backward.

After he had regained his composure he took a closer look. Virtually every one of the menhaden in the net was perfectly formed from its nose to its dorsal fin. But from there back the flesh was missing, as though it been filleted. Nothing showed but skeleton and tail. Mistrusting his eyes, he picked up a fish and stared at the exposed vertebrae and ribs.

In his fifty-two years, James Guthrie had seen nothing like this. And when he tried to imagine what on earth could be responsible, he was left shaking his head.

As for the first person working for the Division of Environmental Management to note that North Carolina's fisheries were being stalked by something never seen before, that would probably be Barry Adams, an environmental chemist in DEM's regional office in "Little" Washington. Fate had clearly chosen Adams to be the point man on this issue. Born in the middle of the state, in the town of Sanford, he'd been the black sheep in his family because his father, uncles, and brother were either wartime or pleasure pilots, but his interest was in the sea. If it could be said that fliers dreamed of sprouting wings, Barry Adams's fantasy was to grow gills. As a kid, he built plastic boats, not airplanes. He fished, he boated; but his interest went beyond recreation to the biology of water: what lived there, and how they throve.

He had planned, after graduating with a degree in biology and chemistry, to continue his studies in physiology, the study of cells. But this was in 1968, and circumstances forced him to trade the ivory tower of academia for the jungles of Vietnam. He survived, but just barely, sustaining RPG (rocket-propelled grenade) wounds in a firefight at an artillery base where the fighting was hand-to-hand. Some vets said they had flashbacks; Adams lived with the memory of Vietnam daily, as shrapnel embedded in his flesh continued to work its way out through the surface of his skin for years after he returned home.

Finding himself unable to go back to school, he took a job with the state, inspecting sewage treatment plants in eastern North Carolina. And while he was at it, he investigated fish kills.

This was an area and a time when things environmental were heating up in North Carolina, and a lot of pressure was put on him to help. Until then, an engineering mind-set, narrowly focused on the solution to pollution as dilution, had dominated DEM. As a biologist, Adams knew it took more than that, but he was one of the few in his agency who considered nutrient uptake and was aware of its relationship to a healthy waterway. So ill equipped was the agency to understand the biological features of the water systems it was supposed to be managing during this period that Adams did not even have a microscope to look through. To identify what was in the water samples he was taking at sewage outfalls and fish kills, he had to borrow a scope from high schools and community colleges. When he'd put in a requisition for one, the response had been, "Why do we have to spend money on this?"

To his mind, the coming of age within DEM, the time when it finally began to recognize the role of nutrient loading, arrived in the late 1970s, with a series of major algae blooms and fish kills on the Chowan River; it just didn't square that what they were seeing could be explained by a lack of oxygen. It took a public outcry and scores of scientific studies before DEM forced the closure of a chemical company that manufactured fertilizer and was dumping enormous amounts of nitrogen into the river.

After the lesson of nutrient loading in the Chowan River, it was surprising to Barry Adams that his agency had so much trouble grasping the concept behind *Pfiesteria:* allow too much phosphorus and nitrogen to be dumped in the water, and you are stimulating expansion of a food web. Which means you are inviting all sorts of new organisms to join the mix, some in small numbers that don't have an impact, some increased to a level that does. Some good, some nasty. It was as simple as that.

It was in the late 1970s and early 1980s that Adams first saw something that confounded him in water samples he collected at fish kills on the Pamlico River. The oxygen levels were fine, salinity and pH fell within the normal range, but when he looked at the samples under the scope, he saw them crawling with microscopic bugs that looked like dinoflagellates. Since neither his training nor his equipment allowed him to go much further on his own, he phoned around the country, describing to the experts what he was seeing.

"Could it be a toxic dinoflagellate?" he asked.

"Impossible. There's no such thing as toxic dinoflagellates in estuarine waters. Only in seawater," they said.

But as certain as the experts were about what it wasn't, no one ever told him what it was.

Adams related it back to the food web. He figured it was a subtle expression of what they had seen on the Chowan River. And he would have been content to let the matter rest there, if he hadn't gone out and collected a water sample on a fish kill in progress around Bath and brought it back to the office, where he set up a test. Filling an aquarium with water, he rigged an aerator to ensure proper dissolved oxygen standards and acceptable temperature, and then he dropped in some *gambusia*, tiny fish similar to guppies, and observed them long enough to feel certain that they were healthy and happy. That became his control tank. Next he scooped out some of the fish with a net, put them in a beaker, and transferred them to another tank in which the pH and DO levels had been adjusted. The only difference was that the water in the tank came from the kill site.

Gambusia were hardy fish. Smashed with a hammer, they would die; otherwise they were tough to kill. But in the killer water, they began to shiver and shake and spiral toward the bottom. In twenty-three seconds they were dead.

Adams was, unknowingly, dealing with *Pfiesteria*. It didn't exhibit the classic shifts he saw with other toxic blooms—didn't turn the water red, didn't cause a jump in the pH—because it was an animal. But he didn't know that at the time. Indeed, one of the few ways he recognized this particular kind of bloom was that when he passed through a fish kill caused by the organism, it was so intense it burned his face, which was why, to protect his skin, he grew a beard.

Back at the office, people thought he was stupid when he tried to tell them, *Hey, there's something out there. There has to be.* But it wasn't until he heard about the work being done by JoAnn Burkholder at North Carolina State, and she came to "Little" Washington and gave a workshop on *Pfiesteria,* that he realized what he had been looking at.

"Thank you, thank you. I've been seeing this for ten years, waiting for someone to tell me what I was looking at," he had said to Burkholder afterward.

Not long before, he had seen *Alien,* and now he thought, *This is just like the creature in the movie. It is constantly changing shape. And when threatened, it has a magnificent defense capability. It can turn into a toxic amoeba, or it can retreat and turn into a rock.*

Adams thought it fitting when he heard that Burkholder had fallen into the research accidentally. That it wasn't really her specialty but she had persisted in the research and finally was the one to make a discovery that had been overlooked by everyone else in the field. It appealed to his Southern rebel sensibilities for an outsider to come along and prove the experts wrong.

At the same time, he feared that the chance of a corporate buy-in at the top of his agency—where those in charge already felt they understood how to manage nutrients in North Carolina's estuaries and had plans in place and goals they were committed to—was going to be remote. And what was so troubling about that for Adams was that if Burkholder's report was true, and this organism was an animal and not a plant, the state's philosophy for dealing with regular algae wasn't going to do when it came to fighting *Pfiesteria.*

As for just how risky this organism was to people, no one knew, but while most everyone he talked with seemed to be satisfied that it was just another germ in a dirty river, which, along with bacteria, could be unhealthy but not really dangerous, Adams had a different opinion.

Starting in the early 1980s, after taking water samples at a series of fish kills, he had incurred lesions on his arms and legs that spread like a rash, creeping up his extremities. His first thought had been that shrapnel was still working its way out, because over the years, metal fragments had surfaced in his skin like rocks in a field. Once, in the late 1970s, as he was putting on a white dress shirt, a sharp piece of aluminum sticking out his forearm literally shredded the sleeve. *My God,* he had thought at the time, *this is like Hook's clock,* a reference to *Peter Pan* and the crocodile that swallowed Captain Hook's watch along with his arm. And just as Hook would hear his timepiece ticking when the crocodile came around, Barry Adams thought of the shrapnel as a haunting reminder that Vietnam was still a part of him.

So at first his focus had been on Vietnam. Even when his doctor told him, "I don't know what this rash is, but rest assured, it's got nothing to do with Vietnam," it hadn't occurred to him yet that it might be connected to the time he'd spent on the rivers sampling and handling dead fish. Nor had he made any associations between the rash and the other problems he was having.

For some unaccountable reason he'd been bothered by an inability to reason logically. He had trouble thinking things through and was uncharacteristically indecisive. It was as if his mind couldn't hold a thought long enough for him to do anything with it.

At the time, he admonished himself for being inattentive, but gradually he realized it was something else. He didn't know how short-term memory was supposed to work, but to him it seemed as though his thoughts were slipping and sliding off the front lobe of his brain.

Tired of forgetting things, he had started living by index cards. But while writing things down on three-by-five cards enabled him to continue to function in the world, it didn't address the other behavioral changes: deep depressions and radical mood swings that estranged him from family and friends; a short temper, which had a lot to do with the dissolution of his marriage; and vicious thoughts that verged on the homicidal. It was clear to those around him that something was wrong, and he knew he wasn't feeling like himself, but he was at a loss as to why.

It was not until he read an account of what happened with Howard Glasgow that Adams figured out what had probably been going on.

Adams suspected that there were a lot of people who had been exposed to *Pfiesteria* and were experiencing similar problems, but because the manifestations were so insidious, affecting the mind more than the body and easily passed off as attributable to something else, just as *Pfiesteria*-caused fish kills had been, it was no wonder that no one realized what the problem was. It took looking back and realizing your judgment was way off on something. Or someone you knew coming up to you and telling you, before you were aware something was wrong, that your behavior was bizarre. If indeed the toxin produced by *Pfiesteria* could have this effect on humans, the person affected would probably be the last to know.

Now that he felt he had a line on what was going on, Adams contacted JoAnn Burkholder. How much exposure did she think was too much, and what precautions was she taking in her lab? He wished to advise others in the field office. With apologies, she said she couldn't be of much help because there was a lot she didn't know herself. She told him he should avoid the kills if he could and, if he couldn't, wear gloves and try to stay upwind. And then she informed him that the health department was in the process of talking with coastal citizens who thought they might be affected, and she asked for his permission to pass along his name.

"Of course," he said, seeing it as a way of collecting more information about the health effects.

Not long after that, he received a call from an epidemiologist and the conversation was a major disappointment.

Maybe the guy was trying not to be an alarmist, Adams thought afterward. *Maybe he was just trying to put me at ease.* But if that was the intent, it certainly failed. The epidemiologist had come across as someone who was doing something only because he had to. The questions were phrased like this: "You haven't had any problems, have you? You don't really think there's a problem here, do you? Smoking and drinking could have caused this, don't you think?" Adams was left feeling that the health official he spoke with did not believe there was anything to this *Pfiesteria* business and was not interested in hearing different.

It was apparent that the man hadn't known he was talking to someone with a sophisticated understanding of proper interview techniques, who also had technical insights into this organism. And what struck Adams as most unfortunate was that he had no doubt that if the man talked the same way to fishermen and crabbers, the information he collected was not going to reflect reality.

Adams thought this was unfortunate, because by this time he had come to consider *Pfiesteria* extremely dangerous. Even without medical knowledge of the internal disruption of organs and processes caused by exposure to the toxin, he found himself thinking about its manifold indirect effects, just as elusive to pin down as the organism was under the microscope. A commercial fisherman who has been exposed to the organism for years and is angry for reasons he doesn't understand goes into a bar for a drink after fishing his pots and ties one on and gets into an argument and blows someone away in a barroom brawl . . . or goes back to an empty house in a state of melancholy and commits suicide. Or a retiree who has moved to a waterfront community in North Carolina is sitting out on his porch, enjoying a southwesterly breeze, with no idea that a fish kill is taking place just offshore, what *Pfiesteria* is, or that its toxin can aerosolize and travel on wind-whipped ocean spray—probably the very worst way to be affected, because direct inhalation into the respiratory system is like an IV going straight into the bloodstream. And so the retiree, whose lung sacs are already damaged from a cigarette habit, finds himself suddenly short of breath but doesn't know what is happening and why he can't breathe. He stands up, gasping for air, takes two steps in the direction of his house, and keels over, dead of a coronary. And the murderer? Gone with the wind.

That one phone call was Barry Adams's first and only contact with the health department. No one ever contacted him again.

He did go to a doctor, who unsurprisingly was unable to establish an etiology for his problems, but after that visit he told his supervisor, "If there's anything to this, and I believe there is, I've been overdosed. And if Howard Glasgow recovered by staying away from the lab, I should keep my distance from the water as well. So I'm here to tell you, I'll hold your hat, but I'm not going on any more fish kills."

That was his final word on the subject. So he never got the chance to tell anyone that he thought *Pfiesteria* was like Captain Hook's clock ticking for the state of North Carolina.

1 6

In early June of 1994, Dr. Ronald Levine sent a letter to the director of Sea Grant in North Carolina, Dr. B.J. Copeland, whose offices were located on the state university's campus. He documented his department's intent to transfer funds to Sea Grant in fiscal 1994–95, for the purpose of engaging in the "vital activity" of investigating reports Dr. JoAnn Burkholder had provided regarding mild to serious adverse health effects related to possible human exposure to *Pfiesteria*. The Sea Grant program, set up by the National Oceanic and Atmospheric Administration to bring the best scientific research to bear on problems facing the country's marine and coastal resources, was a federal-state partnership that distributed grants through some of the nation's premier research universities, and Levine indicated that its history of efficiently directing funds to high-priority research projects of this kind made it the logical choice to administer these moneys. He left no doubt about the high priority he attached to this project: "It is urgent that the circumstances surrounding the substrate/organism/toxin/fish/human interaction be fully characterized as well as the apparent attendant illness."

When she learned that Copeland would be handling the RFP process, Burkholder had mixed feelings. She had received several grants from Sea Grant in the previous five years—with one she had proved that water-column nitrate enrichment could be directly lethal to eelgrass, one of North Carolina's most valuable sea grass habitat species

—and had developed what she thought was a close collegiality with the affable, silver-haired B.J. Copeland, a man with twinkling brown eyes who came across as a country boy but in actuality was a crafty administrator, adept at Southern-style politics, who had been money-brokering the Sea Grant program for more than twenty years.

In many ways, Burkholder thought of Copeland as a mentor—she had turned to him for advice on numerous occasions—so her first thought was that if anyone could make sure the RFP was fairly conducted, it would be B.J. And when she met with him to discuss the situation, he told her, "Don't worry, I'll handle everything. We'll make sure it comes out the way it should."

In order to bring this about, however, he said he needed her help in writing the RFP—"No one knows more about *Pfiesteria* and the important questions that need answering than you do"—and he asked her to come up with a list of a dozen or so questions for the steering committee he would appoint.

The irony of the request wasn't lost on her: the expert was being asked to assist in the design of an RFP that would then be opened to colleagues and competitors. And even though Copeland led her to believe that in the end it would work out favorably for her, she couldn't help but think that this meant the process was going to delay, for several months if not longer, her ability to move rapidly ahead with the research.

As she drafted the questions, it struck her again how difficult it would be for someone who did not have her background and experience with this organism to make much headway studying it. Given how cantankerous it could be in culture, where artificial conditions were imposed in an effort to duplicate a regimen that had been established over time in a natural setting, a researcher attempting to determine what kind of nutrients it responded to, and in what combination and quantity, could easily conduct his experiments in such a way that the dino simply did not respond. It was one of the fears she had about allowing the research to be turned over to those who were unfamiliar with this dinoflagellate. She believed that in addition to the direct nutrient cues that stimulated *Pfiesteria*, there were indirect cues—little triggers thriving in the medium, which had yet to be discovered—and a scientist new to the research could follow an accepted approach for studying dinoflagellates, only to fail because the peculiarities of this organism were not taken into account.

And the concluding opinion would be that her data were wrong.

Even though her research had been widely peer-reviewed by now, there were still those within the academic community who, disregarding the scientific method, rejected her findings not because they had looked at and disagreed with her data but because the biology of the organism was too fantastic, or because when she spoke about *Pfiesteria* it sounded to them as if she were projecting human traits onto this single-cell creature.

She would laugh and say she understood the skeptics' incredulity, because *Pfiesteria*'s transformations were truly phenomenal. She would even admit to times when, watching the organism under the microscope, she was reminded of something out of the *Far Side* cartoon series: the microbe who turns the tables on the inquiring scientist, raising the question "Who's studying whom?"

These people were still thinking of *Pfiesteria* as a tiny plant, which it was not. It was an animal. And the behavioral mechanisms and responses to stimuli it demonstrated were what you would expect from an animal that had developed a range of survival characteristics according to evolutionary interactions. So when she used the term "strategy" to describe its behavior, for instance, it was with full recognition that some people had a problem with its implication of conscious action. But the fact of the matter was, after observation of *Pfiesteria* around prey, it was tempting to use anthropomorphic terms to describe its response.

Burkholder was well aware of the fact that *Pfiesteria* required a new way of looking and thinking, because its behavior ran contrary to most beliefs about dinoflagellates; and that comprehension of its multidimensionality was integral to any successful attempts to answer the questions she drew up for B.J. Copeland. Which was why she was chagrined to learn, when she turned them in, that among those he had selected to sit on the steering committee were two environmental technicians from DEHNR.

Losing all hope that anything good would come of this for her, she walked away from the Sea Grant office that day and told herself not even B.J. Copeland would pull this off; she should forget the half-million dollars slated for her laboratory and resign herself to continuing to beat the bushes and knock on doors in an effort to find a foundation that would give her real support for her research.

At that point she was thoroughly disgusted with the environmental bureaucracy in North Carolina. She had sadly concluded that its policies had to do less with evidence and science than with government officials acting in concert with moneyed interests to preserve the status quo. They said they could not recommend stricter regulations without sound scientific evidence, but they did everything in their power to prevent its achievement. And as if proof of that was required, she learned that as late as the fall of 1994, three years after she had discovered *Pfiesteria* in the Pamlico Estuary, according to official state records it did not exist.

She learned this accidentally, when she was asked by DEM's regional office in "Little" Washington to help them monitor a series of fish kills on the Pamlico because staff shortages did not permit a timely response. So she assisted them, finding high levels of *Pfiesteria* at several sites and afterward turning in a report, several pages in length, that was complete with maps of where the kills occurred, the nutrient concentrations, and the levels of *Pfiesteria* present at the epicenter as well as the edge of the kill zone.

As the regional staffer who had asked for her help scanned the report, he shook his head.

"What's wrong?" she asked.

"Nothing. It's just such a shame that this information, which reflects what was really happening on the river, isn't going to make it into our fish kill database."

When she asked him what he was talking about, he told her that instructions had come down from the central office not to accept as official any information that had not been obtained by its own personnel.

She understood immediately why this was being done. It looked a lot better, and was politically more palatable, for the state, when contacted for information regarding its fish kills, to say that low dissolved oxygen rather than some toxic creature was killing fish. Low levels of oxygen were a natural phenomenon, like salinity and temperature.

They could cause damage to the fisheries but weren't going to hurt people. But it had not occurred to her that the state would engage in subterfuge so blatant, and when she learned of it she simply could not allow it to continue without confronting higher-ups.

She called DEHNR and asked for a meeting, to which she brought a computer printout from the state listing officially recognized fish kills and their causes over the previous three years. And beside a column that read: "DO, DO, DO, Unknown Cause, Unknown Cause, Unknown Cause," she put a page that contained her own database for the same kills; it read: "Dino, Dino, Dino."

"Here is the situation," she said. "According to your records, there is no such thing as *Pfiesteria*. You will find it nowhere, even though your own regional offices have sanctioned me to come and gather data, and agree with me that it has been the prime suspect behind many of the fish kills."

When no one replied, she let out an exasperated sigh and said, "You want to know why I'm critical of the state? It's because of this kind of thing. By refusing to recognize *Pfiesteria* in the official DEM database —and I don't care what you call it: you can call it anything you like— but by refusing to allow it to be identified in your fish kill records, you are not attacking this issue in an honest way. The agency's response is completely irresponsible. In my opinion, this is a deliberate cover-up."

One of the officials finally spoke up. "Well, that's quite an accusation."

Burkholder shook her head, because she knew she was right. No less an authority than the National Oceanic and Atmospheric Administration (or NOAA) had recently identified North Carolina as having among the worst fish kill records in the country.

"It's no accusation. It happens to be the truth."

As for the role of the Department of Health in this "cover-up," she felt it was an active accomplice. As far back as 1989, coastal fishermen had voiced their concern about a cover-up in the investigation of fish kills. The North Carolina Division of Health Services, as the department was known then, had maintained in public meetings that whatever was killing the fish—and it admitted the underlying cause was unknown—was not harmful to people. Their spokesmen attributed the red sores and massive die-offs to an opportunistic fungus, which they referred to as ulcerative mycosis; and they assured everyone that it did not pose a public health hazard. They had even gone so far as to issue statements saying there was no basis to recommend restricting or prohibiting swimming or other activities in waters associated with fish kills.

To Burkholder's way of thinking, the health department's posture in this regard was just as perverse as DEM's. She felt it was a given: the expression of adverse impacts on aquatic communities foreshadowed other biological consequences. And just as the effects of pollution advanced in stages, affecting aquatic grasses first, then shellfish, then finfish, before people started getting sick, so could the pathogenic agent *Pfiesteria* direct itself inexorably toward human beings.

Even if the health department did not want to take her word for what happened to her after working closely with toxic cultures in the laboratory, and even allowing for it to cast a skeptical eye on Howard Glasgow's progressive deterioration after chronic low-level exposures to aerosols from dilute cultures, she felt she had supplied health officials with more than enough circumstantial testimony linking *Pfiesteria* to adverse health effects in the natural environment to obligate them to take the concern seriously. It stood to reason that if bad things happened to people exposed to dilute concentrations of the organism in a laboratory situation, there was a good chance something similar was happening to people exposed chronically to the dinoflagellate in the lower rivers and estuaries.

But as far as Burkholder could tell, the state had mounted a quick and dirty epidemiological survey that was all but guaranteed to be inconclusive and was more about making the problem go away than about determining health risks. If a tree falls in the forest and nobody hears it, then it doesn't make a sound: Burkholder was convinced that this was the operative philosophy of the North Carolina Department of Health when it came to *Pfiesteria*.

It was a formula for disaster, she believed, which was what she had said to Dr. Levine at their first meeting: "You must move on this. We're having fish kills every year. It's only a matter of time before people are hurt too. If you don't act, there will be a disaster."

There was a theory floating around, encouraged by the blockbuster movie *Jurassic Park,* that explained the sudden emergence of *Pfiesteria* this way: Texasgulf was mining a phosphate deposit that went back to the Pleiocene era, when the cysts of this organism reportedly first throve, and a nest of them had been dug up inadvertently and discharged into the Pamlico River.

Burkholder's theory was different. While she was aware that it would be impossible to say when *Pfiesteria* first colonized in North Carolina and whether it was an exotic or a native organism, she believed that it had always been present in the estuaries of North Carolina as a natural part of the fauna, drifting along at a level of activity that was sufficient to sustain it until environmental factors allowed it to flourish in larger numbers. Something in the past ten years had changed the natural ecology of the estuaries, and a threshold had been crossed, encouraging its growth.

Part of what had led her to this conclusion was that even though there was no really good scientific data until recently on the prevalence and severity of the fish kills in the estuaries, when she talked to fishermen and to people who had lived on the coast for a long time, they all said that while there had been kills before, they had never been of the magnitude of the recent ones, nor had evidence of disease been present.

Though she too could only speculate, when she asked herself *Why* Pfiesteria? Why North Carolina? Why now? she felt that a combination of variables had come together to create just the right conditions.

Sometimes it took nothing more than a change in the weather to unleash a killer. In 1993, for example, after an exceptionally mild winter in the American Southwest, there had been an explosion in the region's native field mouse population; consequently more people were exposed to rodents, and some were stricken by a mouse-borne virus that had never been noticed before on this continent: the hanta virus. And in 1983, a strong El Niño brought about a prolonged warming period, and the resultant heating up of the oceans directly affected marine life, setting the stage for global increase in harmful phytoplankton in the 1980s.

At this same time, North Carolina had experienced unprecedented growth in key enterprises that contributed significantly to nutrient pollution. People were moving en masse to the state, generating a lot more sewage and urban drainage. Farmers were increasing their use of fertilizers and pesticides—to the point that a study conducted by the National Oceanic and Atmospheric Administration named Albemarle-Pamlico as the recipient of more pesticides than any other estuary in the coastal United States.

Most of the 127 coastal estuaries in the United States were suffering symptoms of nutrient overload. Nutrients had devastated many of the East Coast's popular fishing spots and shellfish beds. The same with coastal bays in the Gulf of Mexico. Even California, whose estuaries received the nation's heftiest volume of nitrogen and phosphorus, saw its once-productive shellfish industry wiped out by bacteria contamination. But in North Carolina, whose Albemarle-Pamlico estuarine system is poorly flushed and shallow, averaging a depth of six feet, and where there were wide expanses of lush grass beds and algal growths that supported a major fish nursery, it was the compound effect of local conditions and climatic factors that promoted a population boom of *Pfiesteria*.

Although there were some who refused to abandon the belief that Texasgulf was to blame, Burkholder did not consider a single industry responsible for *Pfiesteria*. She was unable even to point to a particular factor, over and above the other contributors, that might have pushed the state over the brink—until the spring of 1995, when she read a series of investigative articles in the *News & Observer*. In a matter of just a few years, it was revealed, a virtual hog revolution had taken place in North Carolina.

She would never forget, and could recite from memory a year later, the opening paragraphs of the first article she read:

"Imagine a city as big as New York suddenly grafted onto North Carolina's coastal plain. Double it.

"Now imagine that this city has no sewage treatment plants. All the wastes from 15 million people are simply flushed into open pits and sprayed onto fields.

"Turn these humans into hogs and you don't have to imagine at all. It's here."

The series made a devastating case that North Carolina had sold its soul and sacrificed its environment to corporate hog farming. It reported how the state, committed to maintaining a semblance of an agriculture industry and seeking alternatives to tobacco, had courted the swine industry with tax breaks, protection from local zoning, and exemptions from tough environmental regulations. It described how an alliance between pork producers and elected officials had transformed the political landscape, allowing laws to be passed that gave hog farms the preferential treatment traditionally extended to family farmers, when the reality was that hog operations were more like factories, with climate-controlled confinement barns and automated feeding equipment that produced hogs in numbers that had attracted international markets and was generating $1 billion a year, propelling North Carolina to the nation's number two spot in hog production, behind Iowa.

The news that a major industry that depended upon lax state laws to maximize its profits had snuck through the back door was sobering to Burkholder, as it was to a lot of readers. But what really got to her were the articles that dealt with the waste generated by these large farms. This vast swine city, as it turned out, was essentially allowed to use the equivalent of outhouses for disposing of ten million tons of waste each year.

According to the wisdom advocated by the industry, hog manure could be handled satisfactorily by flushing it into clay-lined holes in the ground called lagoons, where bacteria were supposed to biodegrade the sewage, which would then be sprayed on nearby croplands as fertilizer. But it was an approach to waste disposal that critics called inadequate and environmentally foolhardy, in light of eastern North Carolina's sandy soils and its shallow water table, which was especially vulnerable to groundwater pollution. Making matters worse, most of the lagoons were conveniently located near waterways; these received the runoff when the surrounding lands couldn't absorb the fertilizer sprayed on them, not to mention the documented cases of farmers deliberately discharging lagoon waste into ditches that led to streams.

The water-quality implications for coastal North Carolina were monumental: Manure that was rich in nitrogen and phosphorus was finding its way into already overloaded "nutrient-sensitive" waters. And as if that weren't enough, the series went on to point out that this expansion had been aided by state agencies that had been slow to act on the growing range of problems resulting from the earlier increase. While state law required all new livestock operations to have certified waste management plans, government officials were leaving certification up to the hog farmers, nor were they sending out inspectors to make sure the plans were being fulfilled or to check for violations. Echoing Burkholder's own conclusions about DEM, the articles declared: "the state's anti-pollution cop has neither the staff nor the will to get the job done."

When the series, which would eventually win a Pulitzer Prize for the *News & Observer*, finished its run, Burkholder was more cynical than ever about the way business was conducted in North Carolina. It was a miserable situation, she thought; one primed to get worse, because leakage from lagoons and intentional releases weren't the only way harmful animal waste could enter the waterways. Sooner or later, she knew, an accident resulting in a major spill was bound to happen.

And when it did, it thrust JoAnn Burkholder back in the middle of controversy.

~~ ~~ ~~

Around seven-thirty on a late-June evening in 1995, the phone rang in Burkholder's office, where she was working late. Rick Dove was the river keeper on the Neuse, one of North Carolina's primary coastal rivers, a job that called for him to monitor the water for signs of pollution. Of all the people whom she had befriended in the course of her fight for the preservation of North Carolina's lower rivers and estuaries, she was closest to Dove, and he would call her whenever there were fish kills or other developments on the coast he thought she should be aware of. But it wasn't the Neuse he was calling about this time. He was phoning to ask if she had heard about a huge swine lagoon rupture on the New River.

"No," she replied. "I've heard nothing. When did it happen?"

"Almost two days ago," he said.

Whereupon she began to swear, thinking that the state was as usual probably holding in confidence information that would embarrass its regulatory agencies.

"How did you hear about it?" she asked.

"A concerned citizen from the area called and wanted to know what was going on. He said DEM had been down there almost immediately, but there hadn't been anything about it in the paper and he was afraid maybe they were hoping no one would find out."

That would be just like them, she thought. But rather than dwell on DEM's shortcomings, she shifted her attention to the rupture's relevance to her own work. For nine months she had been conducting a joint study with Dr. Mike Mallin's lab at UNC–Wilmington to characterize nutrient levels in the New River estuary, where she had previously found *Pfiesteria* in large concentrations, and in order to gather data they had set up a series of sampling stations. From what Dove had said, she suspected the stations were probably downstream from the rupture, which would certainly throw off all their measurements. But then something occurred to her: *I bet nobody is going to properly characterize this spill.*

For months, ever since she'd read the *News & Observer* series, she had been searching for information on lagoon ruptures and what they did to water quality, and she'd found very little. There had been several small spills that the state had cursorily sampled, but the data had been kept in-house and no independent scientists had been involved. Although her suspicions had been aroused, there was nothing she felt she could do about it. Now, however, she recognized how perfect her position was. Her lab had all the baseline data regarding the makeup of the New River, and everything was in place to enable her to characterize and interpret the river after a major spill.

She had meetings she couldn't get out of the next day, so she called her lab and told two of their best technicians, "Folks, this is going to be important. We're going to document just what happens when one of these hog lagoons blows. Gather your gear and get down on the New River. I'll join you as soon as I can."

By the next day, news of the spill had reached the press, making headlines. The dirt wall holding back an eight-acre, twelve-foot-deep hog lagoon at Oceanview Farms in Onslow County had collapsed, releasing twenty-five million gallons of hog feces and urine: that was twice the size of the Valdez oil spill and by far the largest such spill in state history. At the same time local residents were reading that the waste from ten thousand hogs was spreading across farmland, ruining crops and polluting nearby wells, and pouring into tributaries that fed the New River, technicians from Burkholder's lab were rowing up and down a fifteen-mile stretch of the New in a boat, filling sample bottles.

Burkholder arrived the next day, and the sights and smells that greeted her were breathtaking, in all senses. The river was chocolate brown. Bushes lining the banks were filled with dead fish, blown sky-high by the force of the spill. Marinas, docks, and boats were coated with fecal matter. The stench of hog manure in the air was strong enough to make her gag.

The fallout from the hog-waste spill filled the front pages of regional newspapers in the days that followed. A wildlife biologist, who was quoted as saying that much of the headwaters of the New River were dead, estimated the number of fish killed at four thousand, including everything from pickerel and bass to eels and catfish. Investigators looking into the cause of the collapse said it appeared that the level of waste in the lagoon had exceeded the maximum design standard and fines would be levied, an opinion angrily contested by the owners of the hog farm, who said a week of heavy rainfall must have weakened the dam surrounding the waste; they called it "an act of God." Then state health officials, after evaluating the water for fecal coliform bacteria levels and finding that the state standards were exceeded by a factor of 30,000, issued an advisory that the water was contaminated, and swimming and fishing in the New River should be avoided until further notice.

The strongest reaction, however, came from people who lived and worked on the New River; they were outraged at the state's perceived lack of planning, which allowed a spill of this magnitude to occur, as well as confused about the meaning of the health advisory. In the face of mounting demands for an explanation, state officials announced a public meeting in Jacksonville High School to clear the air. In attendance were such top officials as the secretary of DEHNR, Jonathan Howes, and the state health director, Dr. Ronald Levine, and they got an earful from angry residents, who badgered them through the almost three-hour session with shouts, taunts, and accusations.

"Why are you not standing there with your heads held in shame?" demanded a marina owner on the New River, to applause from the crowd. "Have you been on Mars for the past fifteen years? Have you seen these hog operations? Who was the brain surgeon who came up with a lagoon for handling hog crap anyway?"

He wanted to know why restrictions on hog farms weren't sufficient to prevent hog spills, and Howes found it hard to be specific. "This is a circumstance which has never occurred before," he replied lamely. "This is like dealing with the first major oil spill."

Another resident zeroed in on Dr. Levine. "Is it safe for me to be on the river?" he shouted.

Levine too was in a bind: he was unaware of any qualitative or quantitative data on the bacterial or other microbiological flora present in hog-waste lagoons. It had never come up before, so no one in the department had looked into it. The advisory had been issued as a precaution, not as the result of a risk assessment based on hard data. And when his answer seemed to dodge the question, the man cut him off. "Yes or no? Is it safe for me to be on the river in my boat?"

Levine raised his hands. "I'll try and answer your question if you let me finish. My feeling is, it's safe to be *on* the river; it's just not safe to be *in* the river."

Pandemonium broke out. Above the din, someone was heard to say, "Fine. I have my little girl in my boat and we hit a wave and spray gets on her—is she going to be okay? What the hell does it mean it's safe to be *on* the water?"

JoAnn Burkholder was among those sitting in the audience. Prior to the meeting, coastal county health directors had practically begged her to attend. They'd admitted their fears that the true extent of the public health threat was being understated. They said that they had been frantically trying to get state health officials to take more responsibility in this matter, only to encounter a wall of resistance. Fearful for their jobs, afraid to associate themselves publicly with her, given her profile as an antagonist to higher officials in the state environmental agency, they had privately pleaded with her to help them by coming.

At this point in the meeting, Burkholder rose with a question. "I was wondering if the state is going to consider the fecal coliform bacteria contained in sediment in its assessment of when it will be safe for people to return to the water. I'm concerned about this because we have data indicating that while concentrations of coliform bacteria in the water column may be receding, levels are still high in the sediment. Can you give me any indication as to whether you're going to be covering that area before you lift the advisory?" And she gestured to indicate that the question was open to Howes or Levine or anyone else on the stage who cared to respond.

As much as anything, she was trying to make a point: Don't just look at the water; take a look at what has settled on the bottom, where waterborne pathogens (e.g., viruses, bacteria, parasites) survive for long periods of time. She was raising this point because while conducting her independent inquiries, she had come across publications that discussed this as a problem often overlooked, and she didn't want the state to neglect the possibility that bacteriological monitoring of just the water would not give a true indication of how safe the river was for human activity.

Her question had a ripple effect. At the meeting, a health official complimented her on her astuteness and said he'd like to get together with her at a future date, when perhaps she could share her data with him. But the impression that the health department was genuinely concerned about fecal coliform levels in the sediment was dispelled several days later by a physician on staff in Raleigh, someone who had been a valuable inside source, tipping her off from time to time regarding what was going on behind the scenes when he disagreed with the wisdom of actions taken or not taken by his agency, particularly when they related to Burkholder's work. To no one had she revealed that she had a mole in the health department, because she did not want to get him in trouble. But his information had always been reliable, and she trusted him implicitly.

He was calling to say that her concerns were properly placed. "If anything, you've underestimated the potential for problems in the sediment."

"Tell me," she said, "because I know very little about this myself. Only what I've read."

"It's a nightmare," he replied. "I don't even want to think about it."

Then he told her that after the meeting, microbiologists around the state had been asked their opinion of what she had said about the health risks associated with fecal coliform and whether or not its presence in sediment posed ongoing health problems, and every one of them had backed her up. "You're sitting on a powder keg," he quoted one of them as saying.

But when he had relayed this to his superiors, he said, they rejected the information, arguing that the state had no sediment standards for fecal coliform, and besides, there was nothing scientific to prove just how significant a public health problem this represented. He felt demoralized, afraid that the department, pressured to reopen the river, would remove the advisory prematurely, putting the public at risk.

In the weeks that followed, when she was interviewed by the press or asked questions at Marine Fisheries Commission meetings about the New River's safety for human activities, Burkholder continued to speak of sediment fecal coliform bacteria. She would repeat that after a huge amount of livestock waste pours into an area, it takes time for the microbial pathogens to die off naturally, especially in shallow waters such as the New River's, and that any activity that stirred up the river bottom—boating, wading—had the potential to cause problems. And she would add cryptically, refusing to elaborate, that she was not alone in this view; other scientists had informed state officials that they agreed with her data interpretations.

The debate between JoAnn Burkholder and the state Department of Health over health problems created by the hog spill continued throughout the summer. Much as it was apparent that the state was anxious to lift the advisory, it did not, because a water sampling program it put in place on the New River continued to turn up elevated fecal coliform levels.

Internal documents dated during this period indicate that health officials also came to the conclusion that Burkholder's concerns had been justified. One memorandum admitted that fecal coliform was a poor choice for monitoring recreational water quality after a hog spill, other than when "the waste mass has stabilized on the bottom." Another report concluded that the waters affected by the spill did in fact contain a pathogen dangerous to humans, *Vibrio vulnificus*. When it was ingested through contaminated seafoods or infected a wound, the human mortality rate was as high as 33 percent.

But by early September the health department felt that the imminent danger had passed, and Dr. Levine authorized a news release stating that water quality on the New River now fell within the EPA's guidelines; he was recommending that the health advisory be lifted, with the following caveat: "It is prudent to remind the public that all recreational and occupational water activities, regardless of where carried out, are associated with a low but real risk of infection."

When reached by the press for a comment, Burkholder was careful about what she said. For the record she said something about how gratifying it had been to see state officials take the step of issuing a health advisory in the first place. What she kept to herself was her fear that if this wasn't the disaster she'd predicted was coming, then it wasn't far ahead.

And once again she was right.

1 7

In 1987, **Rick Dove** was a colonel in the United States Marine Corps and a general court-martial judge on the bench at Camp Lejeune. His home was at Carolina Pines on the Neuse, one of North Carolina's primary coastal rivers, where on weekends and in the evenings during the summers he and his son would enjoy the pleasures of fishing. They did not fish from a boat, but using two hundred yards of gill net, they would walk out chest-deep in the water, where sometimes they'd stand for two or three hours at a time, pulling fish out of the net, dropping them into containers, and carting them back to shore. They fished for themselves primarily, adding a touch of commercial activity by sometimes selling the crabs they caught out of the back of Dove's pickup truck on the coastal highway. "It was fun for me, a few bucks for him, and everything was neat," Dove would recall. He planned to retire soon and entertained the notion of going into business with his son, running a commercial fishing operation that would supply their own seafood restaurant.

They had a lot of physical contact with the water back then, and there were times, Dove remembers, when they would find themselves surrounded by fish with sores, fish that were dying. Common sense told him something was wrong; but like almost everyone else, he never suspected that what was killing the fish might affect them too. Not even when he and his son developed sores of their own. The state had posted no warnings. No one was saying: Stay out of the water. All he had to go on was his judgment, which he now realizes he didn't exercise properly.

Dove's struggle with short-term-memory loss began that summer, and he noticed it first in the courtroom. A point of law would come up that he wanted to research over the lunch break, and by the time he got back to the courtroom he could not remember why he had gone to the library or what he had looked up. There were even times when he would ask counsel, "What's the topic under discussion, again?"

Dove tried to compensate by taking scrupulous notes. Though concerned, he didn't believe that his abilities as a judge were being compromised. However, when his forgetfulness spread to other areas of his life, he began to worry about a possible brain tumor and decided to consult a professional.

The doctor at the military hospital at Camp Lejeune gave him a simple test. He named ten items, then said, "Repeat back to me how many you can remember in the order I listed them." Dove couldn't get past the last three. Still, a brain tumor was ruled out. Gradually the condition seemed to improve.

It never occurred to Rick Dove that there might be a relationship between his memory loss and the river. After he retired in 1989, he started crabbing full time, only to find that conditions on the river had changed. When he hauled in his pots, with them came a stink off the bottom like raw sewage. As often as not, there would be dead crabs in the pots, and others with holes in them. It was impossible not to recognize that the Neuse River was sick. He also knew he could not in good conscience continue to sell what he caught to the seafood houses when he wasn't willing to eat it himself.

So he went back to practicing law, and he was content doing that for two and a half years, until an advertisement in the local newspaper caught his eye. The Neuse River Foundation, a grassroots environmental organization based in the coastal community of New Bern, was looking for a river keeper, a full-time, paid position.

No one could ever mistake Rick Dove for a tree-hugging do-gooder. He was a self-described Republican and capitalist, with a starchy military bearing, black hair graying at the temples, a deep voice, and a hawkish profile. But like many of those who lived along the Neuse, he felt an almost mystical affection for the river. He knew it had been named for the Neusock Indians, who lived in the area when it was settled by European colonists; that it carried a rich cargo of history; and that its physical majesty, particularly where it widened into an estuary just above New Bern, made it an attraction for tourists and retirees in search of a natural paradise. He also knew by now that troubled waters flowed down the Neuse, and in that advertisement he saw the opportunity to take up the defense of another system he believed in.

Having been an attorney and a judge, Dove understood the meaning of effective advocacy. He knew, for instance, that erroneous information could undermine a good cause faster than anything else. So as soon as he was hired, he set out to educate himself, and relatively quickly he pinpointed the problem. Upstream dischargers were letting nature treat their waste, and they were able to get away with it only because the state agencies whose job it was to control the sources of pollution did not have the political will to tell them to stop or make them pay for the cleanup.

It was while collecting scientific information about nutrient dynamics that he met Dr. JoAnn Burkholder. She impressed him as a woman with grit, integrity, and the courage to stand up for her convictions. She was also extremely knowledgeable. Everything she said about nutrients, fish kills, and water quality made sense.

In the months that followed, Rick Dove came to count on Burkholder as a scientific adviser. And as a way of paying her back, he became her eyes and ears on the Neuse. Since he was on the river constantly, he would notify her when he came across fish kills; and when she was unable to come down to the coast on her own, he would send water samples so she could determine the cause. He considered her a friend and an ally, and when she first began to connect *Pfiesteria* with human health problems, he started to think of himself as a fellow victim.

Now the connection between the sores on the fish and those he and his son had developed in the late 1980s was evident. Howard Glasgow's mental problems reflected his own during his final days on the bench, when he'd had difficulty remembering things. Further, it seemed that he'd improved in the winter months or when he stayed away from the water. Of course, that was too far back to prove anything; but something else supported the idea that he too had been affected by *Pfiesteria*.

Burkholder had told him that physicians suspected that a person could get sensitized to the toxin. "Once you've received a strong dose, it takes very little exposure to bring the symptoms back again," she said. Since he'd assumed the responsibilities of river keeper, there had been days when Dove was out in a boat on the river, monitoring a fish kill, and his tongue would thicken, it would be difficult for him to articulate clearly, and he would suffer afterward from headaches and mental confusion.

Frightening though *Pfiesteria* was—and Dove would be the one to coin "the Loch Neuse Monster"—from his point of view there was one good thing about it. Since taking over as river keeper, he'd found it hard to get people excited about nutrients and algae growth, or about continuing degradation and a system on the brink. Certainly people cared about water quality, but he had come to realize that in order to really motivate people about an environmental issue, information wasn't sufficient: something had to grab them.

To that end he had used every weapon at his disposal. He had helped organize a grassroots air force composed of former military and civilian pilots, who would fly over the river basin, taking dramatic aerial photographs of polluters in the act. It had struck him that people didn't feel as strongly about fish as they did about the wildlife in the forest, and it might be because they had not yet been "Disneyfied"; so he talked a songwriter into producing an album of songs that rendered a fish's private life.

He knew, however, that nothing concentrated the mind quite like human tragedy, and there was some of that with fish kills, because when fish disappeared you were talking about people's livelihood. Not just the fishermen's but whole towns on the coast that depended on money generated by the tourist trade, which was hardly enhanced by harvests of dead fish. Yet fish kills didn't leave a powerful impression unless you actually saw and smelled one for yourself. Even then, even if there were as many as five or ten thousand fish dead, it was difficult for people to know how important that was compared to the millions that were alive.

But a toxic organism that cast a plague upon the waters and threatened to poison people? Now you were dealing with people's fears, with their imagination. And Dove saw that reports of researchers who had

contracted symptoms that could be mistaken for Alzheimer's disease
or multiple sclerosis, simply from breathing the air around a fish kill,
generated a tremendous amount of excitement. All the more so when
it appeared that *Pfiesteria* represented a public health challenge that
the state was either unable or unwilling to confront.

Dove's misgivings about the North Carolina Department of Health
had developed independently of Burkholder's, after a series of events
that he felt could be explained no other way than to say the department
was in denial.

Since assuming the position of river keeper, whenever there was a
fish kill on the Neuse he would call the health department to tell them
fish were dying, they should send someone out to assess the associated
health risks. Sometimes he would get a call back, and sometimes an
official would show up, but nothing ever came of it. It seemed to Dove
that the department's way of determining there wasn't a problem was
not to conduct a study, not to perform tests, but simply to say there
was nothing to worry about.

This had gone on for several years . . . and then the big hog spill
on the New River took place. Curious to see what happened when
twenty-five-million gallons of hog manure was loosed on a river, Dove
took his boat, which he had named the *Lonesome Dove,* down to
Jacksonville, where he put it in the water and motored upstream
to find where the plume began. Burkholder had trained him how to
measure oxygen levels, and when he'd gone far enough to see that the
water was dirty and there were dead fish all around, he dipped his
meter in the water and took a reading. The oxygen level was almost
zero. Then, in an effort to find the forward edge of the spill, he headed
back downstream—and found it within sight of a recreation area in
the town of Jacksonville, where people were swimming, water-skiing,
and jet-skiing. So he went over and asked, "Do you folks know there
was a hog spill upstream two days ago, and now it's right here?" No,
no one knew that. There were no signs posted, they said. No warnings
on the radio or in the newspaper.

After he'd rustled everyone out of the water, Dove called the Onslow
County Health Department. But it was a Sunday, and no one was
working; an answering machine said to call the county hospital if there

was an urgent problem. So he called the county hospital, and that got him nowhere.

The next day, he called Dr. Levine in Raleigh. "Look," he said, "you've got people swimming in the middle of hog waste. There's no signs posted, no warnings, no anything."

Before the day was out, there were signs up at recreation areas and boat landings along the New River, advising people not to go in the water. But Dove knew the health department should have taken the initiative here.

A week later, a huge poultry lagoon ruptured, polluting the Cape Fear River, south of the New, with chicken waste. Again Dove drove down to take a look; this time JoAnn Burkholder had come, to take measurements and collect water samples. The river was too narrow to fool around with a boat, so they drove from bridge to bridge; and while Burkholder went about her business, so did Dove, who had brought his video camera to document their findings. And this was how he ended up with the dramatic footage of a father and his eight-year-old daughter swimming in chicken shit.

With a strong sense of déjà vu, Dove called the Duplin County Health Department. There wasn't even an answering machine to take his call. When he got through to Dr. Levine the next day, he said, "Ron, here we go again. You've got chicken waste. You've got people swimming. I looked at all the recreation areas and boat landings, and guess what? Nothing's posted."

Once again the area was posted the next day, but Dove was left wondering when the Department of Health would have got around to posting either of these rivers if he hadn't called them. And it was the chilling perspective provided by this history, along with the way health officials would soon respond to the summer debacle on the Neuse River, that ultimately convinced the river keeper that when it came to *Pfiesteria*, the Department of Health could not be trusted to protect North Carolina's citizens.

～～～

Except when hurricanes came raging in from the Caribbean, most North Carolinians took it for granted that their sparkling sounds and estuaries, productive fishing grounds and vacation dreamlands, would always be there to be enjoyed. Environmentalists who warned of impending disasters were viewed as scaremongers and extremists. Even when, in the spring of 1995, an unusually high amount of rainfall was followed by a consistently warm and calm regional weather pattern, producing huge algae blooms and masses of submerged aquatic vegetation that clogged the Trent River where it fed into the Neuse at New Bern, the public seemed to be satisfied when state officials said: Aquatic grasses are good. Fish love them. Give it time, and it will clear up.

By the end of June, people were no longer buying that answer. A mat of green scum carpeted the Trent, so solid that it looked as if you could walk on it. So clogged was the river's current that it was virtually unnavigable. There were even reports of dogs dying from drinking the water.

Then strange stories began to emerge from other tributaries to the Neuse around New Bern. Marine scientists studying the effectiveness of an experimental oyster reef in the Neuse found shrimp and clams dead on the river bottom, while "the tops of our reefs were like little enclaves of refugees from war zones. Crabs, worms, anything that could get up on top of them were up there."

All along, the usual suspect had been blamed—excessive nutrients from upstream, creating low oxygen levels downstream—but now angry waterfront home owners began calling local officials and state representatives, saying this was unacceptable and demanding that something be done. Enough pressure was generated that Jonathan Howes, the secretary of DEHNR, agreed to come to New Bern later in the summer for a "site visit" and to meet with concerned citizens.

But long before that happened, events on the Neuse would take another macabre turn: Evenings, the river would be smooth as glass, but daybreak offered the illusion that it had been shattered into a thousand slivers, as its surface glittered with dead menhaden, small croakers, and shad at first; then with mullet, spot, bass, and flounder. On July 24, an environmental specialist from DEM said between 75,000 to 100,000 dead fish were floating along a five-to-eight-mile stretch of the Neuse. Within a week, that number was raised to more than 500,000 over a twenty-five-mile stretch. By the time in-state journalists came on the scene, the Neuse was serving up apocalyptic images. "It looked like the end of the world," was a quote that made headlines.

Rick Dove had seen it coming. Each of the previous summers there had been fish kills, some seasons worse than others, but by July of 1995 he knew this was going to be among the worst. From dawn to dusk he would patrol the river and its tributaries, and he was finding whole creeks that had gone belly-up.

He was also taking a lot of measurements and samples, because in all its public statements DEM was advancing the low-oxygen-level theory as the cause of the kills. While that may have been true in the early part of the summer, by August Dove was noticing gaping red sores on some of the floating fish.

Although he had gone through a number of training sessions with Burkholder, Dove wanted to be sure he was doing things right, so the next time he went out on the river and found himself actually sitting in the middle of a kill, he called her on his cellular phone, describing the physical conditions of the water and reading the oxygen levels on his meter.

"Rick, a loss of oxygen is probably what's happening, but go ahead and take water samples anyway and send them up, and I'll take a look," she said. In the lab, she found the sample teeming with *Pfiesteria.*

Despite the fact that Burkholder identified the presence of the toxic dinoflagellate in lethal levels, DEM continued to maintain that a combination of low oxygen levels and stagnant water was responsible. Before a debate over the cause could heat up, Hurricane Felix settled the matter, rinsing the coast with rain and washing the dead fish out to sea, making it appear, to everyone's relief, that the fish kill "season" had come to an end.

But the drumbeat of bad news for the Neuse had only just begun. In mid-September, there was a fish kill bigger than any Rick Dove had seen in his two and a half years as river keeper. It lasted one week, then two and it didn't seem to be concentrated in any one spot; dead fish were everywhere. Even DEM was impressed, admitting that the numbers soared well into the millions. By the time reporters from around the state converged on the Neuse, the carnage was appalling. The river was "a smelly smorgasbord for dozens of buzzards and other scavengers," wrote the reporter for the *News & Observer.*

In addition to its scale, there were several other differences between this kill and the others that summer: All the instruments showed there was plenty of oxygen in the water to support life. The majority of the dead fish were marked with open bleeding wounds. JoAnn Burkholder and Howard Glasgow confirmed that an active toxic outbreak of *Pfiesteria* was occurring during the kill. And people using the water began to suffer from a weird pattern of illnesses.

~ ~ ~

Over the course of the summer, Rick Dove had noticed how, when a fish kill took place, seagulls would line up by the hundreds on the bridges and fight for a perch on the pilings. Then, at a certain point, cued by something as mysterious as the migratory instinct, they would all take flight, swooping on the dead and dying fish. Of course, there were more fish dead than there were birds to consume them, and it was the leftovers that disturbed Dove. In particular the ones that would wash to shore, where they would be scooped up by poor people standing on the banks with nets. He tried to warn them that the fish with sores might not be fit to eat. And they'd say, Thank you very much. But he never saw them put the fish back in the water.

Dove thought someone should be able to say whether or not fish killed by *Pfiesteria* were safe to eat. At the very least he assumed there must be some plan of action or recommendation concerning public health and *Pfiesteria* fish kills, because this certainly wasn't the first time this issue had come up. Burkholder had been confirming *Pfiesteria* as the cause of fish kills for several years now. So he called a friend in the local health department, who said, "What do you want me to do, Rick? I don't know anything about waterborne health problems."

"Call Ron Levine, then," he said. "Maybe you'll have better luck with him than I've had. But somebody should be out there warning the public. Or at least telling them that they're eating diseased fish at their own risk."

Later that day, his friend got back to him. He had advised Dr. Levine of the situation, and Levine had promised that someone would get in touch with Dove.

But no one from the health department called that day or the next. And Dove grew so frustrated that he decided the only way he was going to get the department to react was to go public. Within the week, three different television newscasts ran a story that pictured Rick Dove piloting his boat up the Neuse, saying, "We need to get some answers from the health department about what to do. Is it safe to use these waters? Is it safe to take fish and eat them? The public has the right to know."

Someone must have been watching, because the very next day Dove got a call from the health department. But it was not the response he expected. On the phone was Dr. Peter Morris, an environmental epidemiologist with the state: the same Dr. Morris who had called him a year earlier to say he was conducting a preliminary inquiry into problems experienced by fishermen as a result of Dr. Burkholder's findings about human health impacts linked to *Pfiesteria* in the laboratory; the very person to whom Dove had given several names, only to learn that he had made fun of these people, suggesting their problems were related to smoking and drinking, not *Pfiesteria*.

After that, Dove had decided he was not going to cooperate with what he perceived to be a flawed inquiry by the state; thinking they were not sincere about trying to find an answer, he had refused to provide any more names of people with mysterious health problems. But now he changed his mind, because the bottom line was that this was the only health department they had, and even though he was disgusted by their apathy, the situation was critical. Whatever happened, he wanted them to be held accountable. He did not want them to be able to say they'd had no idea how serious things were because he refused to cooperate.

When Dr. Morris asked him for the names of all the people he knew who had been out on the river and might have been exposed to *Pfiesteria*, Dove barked out a laugh. "Let me just send you a New Bern phone book, because practically everybody living down here has been exposed to the river at some time or other. And if you're asking for the people who've gone out with me on the boat, you're looking at the wrong people. You need to come down here yourself and speak with the crabbers, the fishermen, the people physically working in the water."

In the end, he gave Morris a short list of names. But he regretted it when he checked back with some of those people and learned that they'd been interviewed by phone and asked if they had sores or memory loss, if they smoke, drank, took drugs. To his knowledge, no one from the state health department came down in person to check things out. No one from the local health department was given instructions on what to do if someone called for information. The public was not advised about how to protect itself. This, when the appropriate response as far as he was concerned would have been to call in a team of experts from the Centers for Disease Control in Atlanta.

With no one else to turn to, people who were experiencing strange symptoms after using the water sought out the river keeper. And during this period Rick Dove became a veritable clearinghouse for personal testimonies about ailments that defied conventional explanation.

He heard from fishermen and crabbers who had not only developed ulcerating sores on their hands and arms after pulling up crab pots from the lower Neuse but were experiencing numbness and tingling; excruciating pain in their joints; and periods of whole-body fatigue and mental disorientation. Many of these people told him they had gone to doctor after doctor, only to be told their complaints were of unknown origin.

He was called by administrators from the YMCA summer camps scattered along the mouth of the Neuse where it emptied into Pamlico Sound, who related anecdotes about odd and persistent infections among staff and kids.

Indeed, the large number of bizarre illnesses reported among children seemed to make a case for acute susceptibility by the young, as dramatized by the account of a ten-year-old who developed a mysterious set of symptoms after spending three long days water-skiing on the Trent River. The boy's mother said it started with a persistent cough that would leave him gasping for breath. His physician couldn't figure out what was causing it but put him on medication, which didn't help. The cough lingered for almost two months, she said, and in that time he developed another problem, which seemed to be related: "He became extremely negative, couldn't see the bright side of things, was hard to please and just plain mean. Friends he used to play with, all of a sudden he lost interest in them. The same with things he used to enjoy and foods he liked to eat. He just wasn't himself at all." She had a public health nursing background—she'd been a school nurse for three years—and said she'd seen nothing like this before with anyone.

In the middle of this outbreak of strange ailments, Dove was visited by the environmental reporter for the *News & Observer,* who seemed skeptical about whether *Pfiesteria* was causing the fish kill, and even to suspect that the health problems might be coincidental. Dove took him on a tour of the river, and as luck would have it, although the river was littered with dead fish, on that day very few had sores on them.

Maneuvering his boat upstream, Dove approached the Neuse River Bridge, where he saw that several Department of Transportation divers were inspecting the concrete pilings. As he came up alongside their boat, one of them commented on all the dead fish. After Dove asked if they'd seen any with sores on them, because he wasn't seeing any, a diver said, "Yeah, we've seen a few," and laughed. "There's plenty of them swimming around down there with holes in them. You just can't see them from up here."

"Can you bring me up a couple?" Dove asked.

After disappearing beneath the surface, the diver came back with one in each hand, both bearing a telltale red sore on their belly.

Dove was holding them so the reporter could take a picture, when the diver volunteered, "You think that's bad, you ought to see the bottom. It's a graveyard of dead fish and fish skeletons."

From there the conversation moved to whether or not the divers had developed any sores.

"Like these, you mean?" One of the divers lifted his sweatshirt to reveal several oozing boils on his stomach.

"Or these?" another man added, showing the same kind of sores on his legs, only bigger.

At a later date, Dove would conduct extensive interviews with the DOT divers. He learned that their line of work put them in a lot of dirty water, the result of pesticides from agricultural runoff, oil spills, illegal dumpings, or dead animal carcasses during hunting season. And they suffered occasional gastrointestinal problems, twenty-four-hour bugs and the like, none of which they could really equate to anything, but they assumed they were job related. For this reason, a year earlier they had upgraded their scuba equipment with a sealed full-face mask equipped with wireless radio communications, and the incidence of minor illnesses had dropped drastically, by almost 90 percent. But that summer, when they started inspecting bridges in the New Bern area, the new system did little to protect them from a set of health problems like none they'd had before: festering sores; neurological problems; mood swings and temper tantrums that created tension with their colleagues and spouses.

With all the publicity surrounding *Pfiesteria* in the press, the divers said they had asked for an update from someone at DEM who was knowledgeable about water quality, and the person who contacted them had laughed off the whole *Pfiesteria* threat. There was nothing to it, he said. Dr. Burkholder's research was sensationalistic. Lack of dissolved oxygen was causing the river's problems.

The divers told Dove that they did not know enough about what was wrong with them to contradict the DEM official, but they had not been completely satisfied with his comments and had asked, "Should we take any special precautions?"

Nothing was said to them about switching from a vest-and-shorts "farmer john" to a full-body dry suit when they were swimming in a fish kill. Nor were they told anything about the possibility of aerosol intake. There was nothing out of the ordinary going on, and standard precautions were sufficient.

Their response had essentially been, Okay, but remember this: We're professional divers. We don't know everything about body physiology, but we do know about the physiology of what happens to us when we're in the water. And these things are not normal.

Rick Dove did not have any answers for the DOT divers. He could not even say with absolute certainty that all the calls he got and all the stories he heard were a direct result of exposure to *Pfiesteria*, because he had no hard proof. But as an attorney, he knew that evidence came in two forms: real and circumstantial. Real evidence about *Pfiesteria* came when you watched it attack fish and human blood cells under a microscope. You could *see* it happen, which made for a strong case. But circumstantial evidence could also be powerful. If the streets were dry when you went to bed at night and wet when you woke up in the morning, you may not have seen it rain but it was a safe bet that it had. And when he took into consideration Burkholder's and Glasgow's personal experiences with *Pfiesteria* in the lab, looked at the pattern of illnesses suffered by people who came in contact with the Neuse River, he had no doubt that this creature was out there, it was doing what Burkholder had hypothesized it was doing, it represented a profound human health risk, and it was just a matter of time before real evidence weighed in to support the circumstantial evidence.

As for why the health department refused to see it this way, he had read in the newspaper what health officials said when asked that very question. "We don't yet have the hard evidence there is a problem here," Dr. Michael Moser, director of the North Carolina Division of Epidemiology, had been quoted as saying. "We have some people who have some symptoms. We have some other people who don't have symptoms."

But Dove had seen nothing that led him to believe the state had seriously gone looking for "hard evidence." Everything he'd seen had led him to conclude that it didn't really want to know if *Pfiesteria* was a real public health threat. And the reason was obvious: If that was proved to be the case—if it was found to cause not only lesions but cognitive impairment and immune suppression and perhaps a host of other serious problems in human beings—then the state had an enormous problem on its hands. Not only from the standpoint of negative publicity—should the Neuse be identified as the Love Canal of North Carolina, it would devastate tourism—but a serious effort by the state to come to terms with it would be politically unpleasant and economically costly.

All that aside, what Dove wanted in the short term was for the health department to warn people away from the Neuse River. And the only reason he could imagine why that had not been done, simply as a precautionary measure, was that it would be perceived as an implicit acknowledgment that *Pfiesteria* was real and was the probable cause of human health problems.

So to highlight the folly of that strategy, he stepped up a public campaign in the press, drawing attention to the uncommon illnesses going around and expressing his deep disappointment in the casual response from health officials, who had a duty "to tell people what they are being exposed to."

At this point the river itself cooperated. If he had been able to write a script for what happened next, Dove felt he could not have outdone fate.

Back in the early summer, when Jonathan Howes of DEHNR promised he would come to New Bern to hear the concerns of waterfront property owners about the vegetative growth on the Trent River, a date of late August had been set for a public meeting. *A nifty feat of timing,* Dove had thought at the time. *By then all the vegetation will have washed away and the furor quieted down.* But the visit had been canceled when Hurricane Felix blew through, and it was reset for the beginning of October, which as it turned out was the peak of the biggest fish kill in North Carolina's history.

The format of the meeting was designed to get scientists and top state officials together in the same room at the same time and give the public a chance to fire questions at them. It took place on Tuesday, October 3, when some four hundred people poured into the Grover C. Fields Middle School auditorium to hear what a thirteen-member panel of "experts" had to say.

While the original purpose of the meeting had been to discuss excessive vegetation in the Trent, it quickly became clear that the question on the audience's mind was: Is *Pfiesteria* harmful to humans who swim, boat, or eat fish caught in the Neuse River? What they got were conflicting answers and an extraordinary glimpse of the palpable tension that existed between JoAnn Burkholder and the North Carolina Department of Health.

Representing the health department in this forum was Dr. Greg Smith, a blond-haired, preppy-looking public health physician who worked in the epidemiological section and appeared to have been dispatched by the department with explicit instructions to get Burkholder. He was combative and baiting; when he tried to be civil it was barbed; and he sparred verbally with Burkholder several times during the long meeting.

When a captain who ran a recreational fishing service asked if it was safe to take sports fishermen out on the Neuse, Burkholder said candidly that no one could answer that question for him because the character of the toxin emitted by this organism was still unknown, nor did anyone know if it accumulated—in fish or people. Those were studies that desperately needed to be done, she added.

Dr. Smith was quick to chime in. Agreeing that "Nobody has all the answers," he launched into a minilecture in which he advanced a series of alternative explanations for the strange happenings that people were linking to *Pfiesteria*. He attributed the sores on fish to bacteria. He insisted that the health effects experienced in a laboratory were radically different from what you could expect in an environmental setting. He reminded people that there were other pathogens in the estuary that were harmful to humans, calling it an inherently "dangerous place." And he finished by pointing to Dr. Burkholder with his pencil and saying, "We need to be sure we don't change *Pfiesteria* into hysteria."

The exchange got testier when, a little later, an elderly New Bern resident told of his grandchild coming down with body lesions after swimming in the Neuse and expressed concern for her health. With a glance askance at Dr. Smith, Burkholder said that not every oozing sore was necessarily caused by *Pfiesteria*, but such sores were a "classic sign" and the ones she'd seen were "dead ringers" for those suffered by workers in her lab. She finished by saying it was precisely because there was a lot more unknown about this organism than was known that, given the toxic outbreak on the Neuse, the state should err on the side of caution and issue a health warning.

With barely concealed sarcasm, Dr. Smith suggested that while they were at it, maybe they should issue one for every stream or duck pond throughout the state, because all of them contained *some* algae. And when that didn't go over very well with the crowd, he switched to another tack, reminding people what all this *Pfiesteria* talk could mean to the business community of New Bern.

"Think of the consequences to tourism," he said.

After Burkholder commented that when the red tide hit Florida that state had acted responsibly, closing the beaches because the dinoflagellate responsible for red tides was known to cause respiratory problems, and the last she heard, the tourism industry in Florida was doing just fine, their war of words escalated. And there was no doubt whom the audience favored. When Burkholder said, "If ever there was a case of Russian roulette," it would be to ignore the public health threat posed by *Pfiesteria*, people broke out in spontaneous applause. As for Dr. Smith, he'd already lost the crowd by the time he committed his fatal gaffe: Called upon to defend his department's dearth of information about health effects associated with hog spills and fish kills, he explained that in the last few years they had run out of travel funds before the end of the year, so they hadn't been able to get around the state to investigate every public health problem that came up.

It was an excruciating moment for the state officials on the stage, painfully so for the secretary of DEHNR, who was sitting between Dr. Burkholder and Dr. Smith, and who at times had looked as though he were caught in a crossfire. It was obvious that this was news to Jonathan Howes. What was so shocking about the statement was that if it was true, it was a shameful acknowledgment that the department was unable to fulfill its public health responsibilities for budgetary reasons. If it was not true, why not offer it as an explanation? Either way, it was an indictment against the credibility of the health department on the *Pfiesteria* issue.

If Dr. Smith had any notion that his remarks about tourism would calm matters that night, he was soon disabused. A columnist for the *News & Observer* gave his warning a deserved lampooning: "The whole thing reminds me of the movie *Jaws*. The good tourists of Amity were getting munched left and right by a Great White Shark. But Amity's cheesy chump of a mayor refused to close the beaches or warn people of the danger they faced. He just couldn't lose sight of the almighty dollar. OK, it was a movie. And people aren't being swallowed whole down there in New Bern, but the principle is the same: Citizens' health remains at risk while officials hem and haw."

But by the time that column went to print, more bad news had come off the Neuse.

The morning after the meeting in New Bern, Howes had agreed to allow Rick Dove to take him on a boat tour of the river. Before that, it had seemed to Dove that DEHNR's entire staff were in lockstep in their belief that reports from the coast were overblown, water quality was not as big a problem as citizens were making it out to be, and it was all part of an environmentalist plot to stir people up. By denying there was an emergency and refusing to take protective or remedial action, he felt, the state had turned the average citizen into an activist, which explained what had happened at the meeting the night before. When he'd heard the argument that had taken place between Smith and Burkholder, it had been all he could do to not get up and drag Smith down to the river, saying, "You're telling me you didn't have gas money to come down and look at this? Well, look. *Look.*"

But for Dove, that meeting would stand as the last failed attempt by the people in control of the state environmental and health bureaucra-cies to put the best face on a wretched situation, because once Howes was out on the river and able to see with his own eyes the magnitude of the mess, the man had no choice but to recognize that it would be pure stupidity to try to maintain an everything's-okay position.

Several years earlier, the DEHNR secretary and Dove had had a conversation about the deterioration of the Neuse, and at that time Howes had told him, "Rick, you've got to understand the political realities at work here. The hog guys, the developers, they've got the governor's ear, and he's going to do what they say unless somehow you can figure out a way to put big-time pressure on him. You've got to get people to start screaming and hollering, because until you do, I can't do anything." When they came in off the river and were sitting in the office of the Neuse River Foundation, Dove recalled that conversation and said, "Jonathan, once you told me we needed to holler. We did. Now it's your turn."

Howes asked if he could use the phone, and Dove left the room. But from what he was able to overhear, he gathered that Howes was calling both Governor Hunt and Dr. Levine and recommending that they take action.

Two days later, state authorities declared a Health Warning for the Lower Neuse River. People were told not to swim, fish, or come in contact with waters where the toxic dinoflagellate *Pfiesteria* was killing fish, or to consume any part of a fish with sores on it. It was the first pronouncement of its type ever issued in North Carolina; before this, such warnings had been issued only when there was scientific data confirming the existence of a problem, as with dioxin or mercury contamination. Even though the state lacked the technical data to prove a health risk in this case, by issuing a warning it was formalizing the concern.

For the river keeper, it was a historic step, significant on both a symbolic and a real level. Certainly he felt the state could and should have responded a lot more quickly. But what was memorable about this, in his opinion, was that for the first time North Carolina was admitting that its policies for protecting coastal waters were a failure. And that was no small thing. That very summer, in the middle of the state's worst fish kill, he'd heard Steve Tedder of the Water Quality Section stand up at a meeting and brag about what great progress the state was making in cleaning up its lower rivers and estuaries and how North Carolina was the envy of other states. Dove had had to struggle to control his anger, because he knew it was untrue and this was another example of image over substance, a recurring theme with the state. Officials of DEM seemed constitutionally incapable of acknowledging that any of its waterways were broken—until now. And the importance of this warning, in Rick Dove's mind, was that a person had to admit something was broken before he could seriously go about getting it fixed.

Local reaction to the advisory was mixed. The national news media had picked up on the story, and people around the country were reading about these events and seeing them on TV. Not everyone in the New Bern area appreciated the coverage. Many commercial fishermen, for example, feared that a public panic over the fish kills could devastate the local seafood industry. That concern had a legal corollary. Many fish and crabs taken in the Neuse River and Pamlico Sound were sold to markets for interstate trade, and as yet there had been no studies done to determine whether it was safe to cook and eat fish that had been exposed to *Pfiesteria*'s toxin, theoretically opening the state to personal liability should it ever be determined that the food was tainted.

Within a week, the North Carolina Marine Fisheries Commission issued its own proclamation, closing a ten-mile stretch of the Neuse to both commercial and recreational fishing, including shellfishing. Once again it was not an action taken on the basis of evidence that the seafood was contaminated; rather, it was undertaken to address a perception problem and to maintain the public's confidence in North Carolina's fishing industry. But qualifications and clarifications aside, it was one more sign of failure.

As pleased as Rick Dove was to have public officials finally accepting what he and others had been saying for years, it was no cause for celebration. "My river is still dying," the river keeper was quoted as saying as late as the fall of 1995. And it scared him to think, if *Pfiesteria* was as injurious as he believed it to be, how many people had been imperiled from eating fish infected with its toxin, how many divers swimming in infested waters had absorbed its toxin, how many kids water-skiing and playing in the river had inhaled its toxic vapors, all because these warnings had not been issued earlier. He was afraid that the full danger of *Pfiesteria* had yet to be experienced, and that in terms of the bad news about the river and health effects, the worst was yet to come.

In the wake of the New River hog spill, pork producers, angry because they felt Burkholder was singling them out unfairly as the main polluters of North Carolina's lower rivers, went on the offensive. Mindful of N.C. State's strong financial ties to agriculture, they tried to get to her through her university. The dean of her college was asked to tell her to tone it down, she was blowing the "hog problem" out of proportion. Besides, a scientist's job was to provide objective information, not criticize an industry that had been economically good to North Carolina.

Whenever they had a chance to comment, health officials called her credentials into question, pointing out that she was neither an epidemiologist nor a licensed physician, and that when she contradicted the opinions of those who were both, she was straying irresponsibly outside her area of expertise.

And these were two of the more *civilized* responses! Following the meeting in Jacksonville, Burkholder received two anonymous death threats. "Bad things happen to bitches who don't know how to keep their mouths shut," a deep male voice growled when she answered the phone around eleven o'clock one night. The other call was along the same lines: "People who are too stupid to know when they should shut up sometimes wind up dead."

Then, at a meeting between Onslow County commissioners and state environmental officials to discuss the condition of the New River,

Commissioner Sam Hewitt complained that Burkholder's sediment reports and work with *Pfiesteria* was "scaring people to death." He suggested the state had enough problems without her and called upon state officials to find a way to silence her and scientists like her. "I'd like to take a rubber hose to some of them," he declared.

Because the potential stakes were so high, this was a situation in which a lesser scientist might have been tempted to resort to hyperbole. But Burkholder remained levelheaded and was careful not to push her claims beyond the bounds of available evidence or scientific proof. Yes, the health hazards associated with animal waste were still poorly understood, she replied, but this much was known: almost one hundred microbial pathogens harmful to humans could be found in livestock waste, some of them had a greater infectious potential as a result of the routine use of antibiotics in animal feed, and the state's standard, which was designed to protect humans, tested for very few of them. As for her credentials, she never pretended to be an M.D., nor had she claimed to be an epidemiologist. But she was an aquatic scientist; she had received graduate-level training in microbiology and taught a graduate-level class in food, milk, and water sanitation; and her research into water-quality issues had given her an extensive education in fecal coliform problems. And to her anonymous callers, all she had to say was what she shot back at them when they tried to intimidate her on the phone: "Fuck you."

In the middle of all this had come the big fish kill on the Neuse, accompanied by a rash of illnesses that, in her mind, represented a perfect example of an emerging disease phenomenon.

Clearly, coastal North Carolina had suffered an environmental calamity, this time not from a hurricane coming in off the ocean but from pollution pouring down its rivers. At least that was the message the media broadcast round the country and indeed the world. In some cases it was difficult to tell whether a press that ran with headlines such as "Will Playground Turn to Killing Ground?" was reporting on the events that occurred or promoting the worst-case scenarios. If an award were given for exaggerated comparisons, the *News & Record* of Greensboro, North Carolina, would have won hands down: "The story of *Pfiesteria* is part science fiction, part murder mystery. Think of

Alfred Hitchcock's *The Birds,* where normally meek warblers turn homicidal. Think of Jack Finney's *Invasion of the Body Snatchers,* where alien 'pods' take over human beings and render them zombies."

But as creative as the press was in reporting the low points of the summer of 1995, to its credit it was equally outspoken in the way it took the editorial high ground, drawing attention to the reasons *Pfiesteria* found life sweet in the Neuse—"it contains all manner of fertilizer, human and animal wastes, chemicals and who-knows-what-all"—and needling officials who had put off as long as possible the decision to tell folks it was probably wise to stay out of the Neuse River, and no, please don't eat fish from it.

The Wilmington *Morning Star* knew exactly what had taken so long: "The Hunt administration didn't want to scare off any of those Yankees who might want to visit the pretty houses in New Bern and stay overnight and eat some fish and buy some souvenirs." It also knew how to solve the pollution problem. "About the only thing likely to get the state to clean up our waters is for a few tourists to go back home with fevers and sores and then gripe to the national news media. Boy, you just can't buy advertising like that."

The fear of "advertising like that" so shook up the travel and tourism industry in North Carolina that they sent a letter to Governor Hunt and key representatives of the General Assembly, reminding them that travel and tourism was the state's second largest industry, that its economic impact exceeded $8 billion annually, that those who chose North Carolina as a travel destination did so because of its world-renowned beauty and the quality of its natural environment, and that "If we spoil it, they will not come." There remained the specter of the red tide of 1987, when the dinoflagellate *Gymnodinium breve* devastated the coastal economy, not only shutting down harvesters, processors, retailers, and distributors in the fishing industry but taking a billion-dollar cut out of tourism markets such as beach resorts and hotels and restaurants.

In the face of all this negative publicity, as well as in response to howls from not only environmentalists and fishermen but also home owners, real estate agents, and mortgage bankers, the state's political leaders announced that they'd seen the light and promised to take swift remedial action. DEHNR secretary Howes called for a reexamination of the state's approach to dealing with nutrients in the Neuse River. In the legislature, a bipartisan committee on river-water quality and fish kills was formed.

Of more pressing concern was how to reestablish public confidence in North Carolina seafood after the issuance of a health advisory and a ban on fishing in a section of the Neuse. And toward showing the rest of the country there was nothing to worry about, as soon as the fish kill was deemed officially over DEHNR organized a big fish fry. On a sunny Tuesday, in a picnic-like atmosphere at Union Point Park on the shores of the Neuse in New Bern, platters of freshly caught flounder were served to smiling public officials, who ate heartily while posing in front of the TV cameras.

It was an awkward scene. The turnout was sparse, with as many members of the media present as there were city, county, and state representatives. Rumors abounded that the fish that were served came not from the Neuse River or Pamlico Sound but from the Atlantic Ocean; and for many, the spectacle had a reverse effect, playing into the public paranoia, especially when news reports made light of the event as an effort to "take North Carolina seafood out of the fire and put it into the frying pan." But as an assistant secretary from DEHNR candidly admitted afterward, it was a no-win situation. The state had to do *something* to save North Carolina's seafood industry.

Meanwhile more reports of mysterious illnesses were coming in, not just from residents of North Carolina but from people who had visited the state over the summer. Calls were also fielded by chambers of commerce from people around the country, asking if it was safe to plan a vacation to North Carolina or to send their kids to a YMCA camp there the following summer.

On days when you would expect to see the Neuse crowded with sailboats, there were none. The river was as empty as a sea.

Throughout this period the health department continued to downplay the notion of a health hazard from *Pfiesteria*. They emphasized the difference between "self-reported data" and "scientific data," of which there was not enough available to prove a "cause-and-effect relationship" with any of the health problems. But given the context of an estimated fourteen million dead fish, the distinction provided little reassurance.

It was in this highly charged atmosphere that North Carolina Sea Grant director Dr. B.J. Copeland announced that the state had approved $550,000 in research funds to answer questions that everyone was asking about *Pfiesteria,* and the recipient of the lion's share would be Dr. JoAnn Burkholder.

There was only one problem. If Burkholder was the big winner, why was she so unhappy?

After helping B.J. Copeland develop a list of questions, Burkholder had done her best to forget about the *Pfiesteria* funding. She was aware the process seemed to be moving slowly, but she hadn't given it a lot of thought until she happened to be in the Sea Grant office, talking with Copeland about another matter, in December of 1994, and almost in passing he mentioned that DEHNR had divided the *Pfiesteria* funding into two components—environmental and health—and that researchers from East Carolina University had been selected to receive money out of the health component for an epidemiological study.

At the time, she had said nothing but thought, *That must have been the most quietly conducted RFP in history.* Afterward, she found herself thinking that something odd was going on. If the epidemiological component had been decided upon before a Request For Proposal was issued, it violated protocol.

A month later, she was given another reason to question the integrity of the RFP process that Sea Grant was supposed to administer. Conversing with a scientist from the National Marine Fisheries Service, Dr. Patricia Tester, who said Copeland had asked her to sit on the panel reviewing the *Pfiesteria* proposals, Burkholder mentioned that she'd been told the money for an epidemiological study had already been apportioned and asked how that had happened. Tester responded, "What are you talking about? I don't know anything about that." Then, several months later, Burkholder had another conversation with Dr. Tester, who this time told her that she had resigned from the grant selection committee because she had too many questions about the way B.J. Copeland was handling the DEHNR money.

This was a major cause for concern, because the RFP process was supposed to abide by strictly defined procedures and follow a clear standard of ethics. It was important to the credibility of scientific research that the process respect its own rules.

For this reason, when a packet with the RFP announcement arrived in Burkholder's campus mailbox in late August of 1995, she skimmed the covering letter and determined that she had six weeks to submit a proposal, then tossed it aside. She didn't read it closely because she didn't want to think about it, and because she was busy documenting the role of *Pfiesteria* in the ongoing fish kills.

But she did not forget it altogether, and two weeks before the deadline, she retrieved the packet and started to read it through. She was familiar with the Background section, because she had provided most of the language for it. The same went for Objectives and Proposal Guidelines. When she arrived at Proposal Limits, however, she was stopped by point number one, which said that proposals would be limited to $75,000 per investigator.

She picked up the phone and dialed the Sea Grant number. When Copeland came on the line, it was apparent that he'd been expecting her to call.

She made an effort to be polite but knew a curtness came through, because one of the fears she had expressed in a letter to the governor a year and a half earlier was that the funding would be piecemeal, so that the resultant research would be effectively diluted. She had predicted this would happen and pleaded with the governor to make sure it didn't. She had even shown Copeland a copy of that letter.

"Why was a ceiling of only $75,000 put on what each researcher could receive?" she wanted to know.

"Well, the committee was very firm on that point, JoAnn," Copeland said, explaining that they did not want to give all the money to just one researcher. "Everyone agreed that a multidisciplinary approach would be best." He went on to tell her that this didn't mean she could not submit a multiple-investigator proposal; it was just that they were limiting the amount a single investigator could receive.

She thought, *The objective of the program should be to fund not as many people as possible but the best people to answer the most important questions.* But for the moment she let it pass.

"I also see that there is no money for equipment," she said.

This too went back to one of their earlier conversations, at a time when it had been her understanding that Copeland was working *with* her to shape the *Pfiesteria* proposal. As it stood, she and Howard Glasgow were the only two people in the state of North Carolina who could identify *Pfiesteria* in all its stages. When the state wanted to determine whether *Pfiesteria* was present in a water sample, it relied on them to identify the organism; and in the past several years she had provided many sample analyses to the state free of charge and conducted several workshops in which she had tried to show state biologists how they could identify it on their own. But the organism was so polymorphic that her mediocre microscope hadn't been much help. The state was still sending her samples, and if they could upgrade her microscope, the sharp pictures it could provide would constitute a pictorial atlas for future reference.

Another item she had thought was important, and had listed on her preproposal to Dr. Levine, was a backup generator for the dinoflagellate facility, so that if there was an unexpected power outage, the cultures would not be destroyed.

Sounding like a broken record, Copeland said, "Well, the committee was firm on that point too. No equipment."

At this point she returned to the $75,000 limit and what it meant to the research she wanted to do regarding the ecological component: how the organism was activated and stimulated; the biological triggers that caused it to produce a toxin; and the means of controlling its growth. It was the research she'd focused on for several years and had originally expressed a desire to pursue, and she couldn't imagine how she was going to be able to accomplish that on so little money.

Copeland told her she didn't have to worry: "The committee is looking for and expecting you to provide toxin so we can get it characterized."

That Copeland was expecting her to provide toxin did not come as a surprise. All along she had said that once the mandatory upgrade of her dinoflagellate trailer to a Level 3 biohazard facility was completed, she would begin culturing the organism in the quantities necessary for toxin characterization. But if all she could expect to receive from this grant was $75,000 to produce toxin, this left her nothing for research.

Her throat was oddly dry when she spoke. "This stinks, B.J. I don't want to do it. I don't want to go through this."

"You need to do it, JoAnn. You've come this far."

"But this doesn't allow me to do anything on the nutritional ecology, which is what I wanted to do."

"JoAnn, after all this, what's it going to look like if you don't cooperate?" He answered for her. "It's going to look like sour grapes. Like if you can't have all the money, if you can't do what you want, then you won't play."

She took a deep breath to settle her emotions.

"JoAnn. If you really are interested and care about what is happening to people in this state, you will do this."

She thought it highly unusual for the director of a granting agency to release an RFP with bullets listing the research objectives and then, instead of letting the investigators choose, to tell one of the prospective applicants: Here's what the steering committee expects you to be responsible for. And in a follow-up conversation, Copeland gave her further "guidance," encouraging her to include as part of her proposal coprincipal investigators who would address some of the other pressing questions, such as the chemical characterization of *Pfiesteria*'s toxins, and a small-mammal study to provide insights about potential health impacts that would be more analogous to humans than tests with fish.

Of course she recognized the importance of these issues, but they were not in her area of expertise. What she'd always wanted was to verify what level of anthropogenic nutrient enrichment stimulated *Pfiesteria* to grow in higher abundance, in order to help design controls that would discourage its growth.

She did not ask Copeland who, if not she, was going to do the nutritional ecology component, because by this time he was showing his irritation over the fact that she was not graciously accepting his counsel. But it was all so questionable for her and so different from the fair trial he had promised that she was no longer sure of the right thing to do.

And then she received an unexpected phone call that put everything into perspective. The caller identified himself as Dr. Aaron Schecter, and he was calling to request copies of all her publications on *Pfiesteria*. He didn't seem to want to say much more than that, which struck her as strange, so she kept him talking, and as the conversation unfolded, it turned out that he was the epidemiologist at East Carolina University who, over a year earlier, had been assured by the Department of Health that he would be receiving a sizable portion of that money to conduct an epidemiology study.

They went on to talk for almost two hours, during which Schecter leveled with her, telling her that he too was in the process of writing an RFP but that he'd been told it was simply a formality and he could expect to receive funding when it was complete. He made several other distressing admissions. When Burkholder asked why it had taken so long for him to get in touch with her, he replied that he had been advised not to.

"Why not?"

Because, he was told, she would try to poison what they were trying to do.

She thought about that a minute. And knowing that she alone had maps that identified the locations of toxic outbreaks, which were of obvious value to someone trying to study the effects of toxic exposure, she asked, "Tell me: if you weren't supposed to get in touch with me, how were you going to be able to do an epidemiology study related to *Pfiesteria?*"

"Actually, we weren't. The people at the health department said there was not enough reliable data on that, so there wasn't going to be any way to establish a relationship. This is going to be an occupational health study of crabbers."

It took her a moment to swallow that, because from the very start, everyone had agreed a major epidemiology study was needed.

Almost afraid of the answer, she asked, "Well, if you wanted basic ecological information about this organism, where were you supposed to go?"

"We were told if we wanted to obtain any advice about *Pfiesteria,* we should consult Dr. Hans Paerl, because from now on he was going to be *the* credible scientist identified to work with this organism."

She closed her eyes. Hans Paerl was a highly respected scientist at the UNC Institute of Marine Science in Morehead City, known for his nutrient studies on blue-green algae in freshwater rivers. Within the marine science community, it was said he "owned" the Neuse, which was why, she'd been told by people who worked in his lab, Paerl had been jealous that she had discovered *Pfiesteria* under his nose. For a long time he had refused to believe it even existed. Though he finally came around on that point, he still refused to acknowledge its role as a fish-killer of significance. In scientific papers and in interviews with the press, he had continued to maintain that low dissolved oxygen due to nutrient overload was the primary fish-killer in the Neuse and that *Pfiesteria* was a secondary and relatively insignificant opportunistic organism. "Nothing more than a sideshow," was the way he referred to it.

This was, understandably, a position that state officials had welcomed. And if what Schecter was saying about Paerl was true, and if Paerl remained true to his previous position and his findings were much different from hers—if, for instance, he was unable to establish that the organism was stimulated by nutrient loading—then the state would be given a great excuse for doing nothing. It was the old dodge of pitting two opposing scientists against each other so officials could say: Look, even the scientific experts can't agree.

Her conversation with Schecter took place ten days before the grant applications were due, and when she hung up she marched straight to Copeland's office with what she'd been told.

"What is this?" she demanded to know.

"I don't know anything about it," he said. "I don't know where those people could have gotten that impression."

The hair on her arms stood up, literally. Because she felt from the way his eyes drifted away from her that the director of Sea Grant had sold her out.

Feeling, as she would say later, "as if a gun had been placed to my head," she threw together a proposal titled "Continuing Effort to Characterize Impacts of the Toxic Estuarine Dinoflagellate *Pfiesteria piscicida*," in which she proposed to culture toxic stages of the organism in mass quantities in her laboratory and provide toxin to a team of investigators she had assembled. Among them were an organic chemist—Dr. Francis Schmitz—at the University of Oklahoma for the purposes of toxin analysis, and an environmental toxicologist—Dr. Ed Levin—at Duke University for neurobehavioral studies on rats, after which a professor of medicine and neurobiology, also at Duke—Dr. Donald Schmechel—would conduct postmortem examinations of the rat tissues. And then, because she simply could not bring herself to submit a proposal that did not include something about nutritional studies, she tacked on a small request in that area, asking for $33,000 and listing Dr. Mike Mallin as a coinvestigator.

For the next month and a half she heard nothing. Just before Thanksgiving, Copeland called her into his office on another matter, and while she was there he told her that a decision had been made on the grant funding and she was in, her proposal was going to be funded. But he also confirmed what she'd already suspected: researchers at East Carolina had received the epidemiology component, and Dr. Hans Paerl was being given the nutritional ecology experiments.

Her conversation with B.J. Copeland lasted for one hour that afternoon, and it solidified her fear that the promise of a fairly conducted RFP was a travesty. Not only had a flawed epidemiological study been funded, the nutritional work was being handed over to a rival who had publicly expressed skepticism about her research—while she, the researcher who had discovered the organism, the scientist who was universally recognized as the *Pfiesteria* expert, had been relegated to the status of a technician. It meant that her lab would be sacrificing live fish seven days a week around the clock, enslaved to the dirty and dangerous work of producing toxin for other researchers.

Copeland tried to deny it. He said the epidemiology component had been competed for fairly. He said that Paerl's proposal had received the highest reviews of anyone, that in fact she had received two negative reviews, and that Hans Paerl was more experienced than she was. To which she rebutted that she had twenty years of experience in this area, that Paerl had never worked with dinoflagellates before, had not published a single paper on fish disease, and didn't even know how to identify *Pfiesteria*.

"B.J., you are helping people who don't know what they're doing, and I don't understand why," she said in summation.

But arguing at this stage was pointless, and she left their meeting feeling angry and betrayed.

When the results of the RFP were announced and the newspapers reported that epidemiologists from ECU received $126,000, Hans Paerl and a coinvestigator got $113,000, and JoAnn Burkholder was awarded $319,000, people called to congratulate her, because it appeared that she had received generous funding. What no one realized was that her proposal had been customized to what the Sea Grant director wanted, so most of the money was going *through,* not *to* her; that her special area of research had been given to someone else; and that she had ended up with next to nothing for research.

I'd love to be wrong just one goddamn time about these people, she thought when she read the news.

Over the course of the next few months, Burkholder met several times to discuss details with Copeland, and he told her that in her proposal she had asked for over $400,000, more money than the committee had allotted to her, so she would have to cut it by one-third. He said he would not be able to complete the funding process without a revised project budget; but given the strong interest in the mammalian studies, he would go ahead and make a direct award to her coprincipal investigator at Duke, so he could get started. She agreed to begin providing Dr. Levin with toxin as soon as it could be purified.

But she was not inclined to hurry up and submit a revised budget, because she had learned of yet another irregularity in the funding process. Dr. Paerl had acknowledged in his proposal that he did not know how to identify *Pfiesteria,* but he covered himself by stating that he would be assisted in his research by a dinoflagellate expert from the University of South Carolina, Dr. Richard Zingmark. But in March of 1996, Zingmark stated in a letter to Burkholder that he did not know how to identify this organism either, and he wondered if she would be willing to train him.

She began to give serious thought to rejecting the money, but two things held her back. She knew that if she rejected it, she would be characterized as obstructing the process that was intended to provide an understanding about what people needed to know to protect themselves from *Pfiesteria,* and that bothered her greatly. And she was afraid that if she turned the funding down, it would be given to someone else and she would be eliminated from the *Pfiesteria* picture completely, which she had come to suspect was probably the intention underlying all these machinations.

There was a brief moment when she thought she saw an alternative. Officials at the Environmental Protection Agency were interested enough in her research to fly her to Washington for a meeting, and she came back hopeful that something would come of it—only to receive a letter saying an emergency had been declared within the agency, all their funding was frozen, and they were sorry but they were unable to consider additional projects at this time.

It was an impossible situation, and when she turned for advice to friends who were aware of what was happening—"What would you do?" she would ask them—most said that she appeared outflanked and if she wanted to maintain a stake in the research, she should accept the funding.

In the meantime B.J. Copeland was calling and pressuring her to make cuts in her original proposal and submit a revised budget. But she continued to stall him, because she did not want to make it easy for him to get away with this. She intended to string him along as long as she could, hoping for something to happen that would help her make a decision, hoping he would eventually make a stupid mistake.

When it happened, it was so unexpected that it caught her off guard. In early March, she was attending a conference in Morehead City dealing with coastal issues, and at a break in the proceedings she was approached by a dark-haired scientist. He said his name was David Green, that he worked at the N.C. State Seafood Lab in Morehead City, and that he had been hoping he would see her there, because he and his collaborators were the ones who were doing the study to determine whether fish exposed to *Pfiesteria*'s toxin were safe to eat.

She looked at him. "This is the first I've heard about that."

He seemed slightly surprised, but continued as if it didn't really matter. "Well, we've been told that you would be providing us with toxin, so I'd like to tell you more about what we're doing and see when we can arrange to pick up some of the toxin."

Knowing nothing about this person or his project, she merely said, "Fine, give me a call."

Several days later, she received a message on her answering machine from a woman who identified herself as Pat McClellan Green. She worked at the Duke Marine Lab in Beaufort, where she and her husband, David, were collaborating on the fish-toxicity study. Burkholder returned her call, and a very interesting conversation ensued. Dr. McClellan Green said that at a banquet hosted by the North Carolina Fisheries Association in November 1995, B. J. Copeland had spoken to them about the feasibility of conducting a study to determine whether fish exposed to *Pfiesteria* were safe to eat. He had said there was money available through the *Pfiesteria* grant, and that he'd see to it that Dr. JoAnn Burkholder would supply whatever amount of toxin they needed. So they put a proposal together—

"Wait a minute, wait a minute," Burkholder said. "This is in November?"

"Yes."

"And B.J. is telling you that I would supply you with all the toxin you needed?"

"Yes."

She thought about that. "Why didn't you call and ask me if you would be able to rely on my services?"

"I didn't think we had to. B.J. made it sound like it wasn't necessary."

She thought about that too, getting it straight in her head. These people had not submitted a proposal until the month after the deadline, at which time Copeland had indicated that he would be willing to fund a project with them using *Pfiesteria* money. Furthermore, he had told them, without talking to her first, that she'd give them what they wanted, free of charge.

From what Burkholder could tell, Dr. McClellan Green seemed genuinely unaware of the history behind her request. So she said, "Look, before I can proceed with this, I need it in writing. And please, include in your letter what brought you to this point, and tell me how much toxin you think you're going to need."

When she hung up, Burkholder found herself wondering if she had dreamed the conversation. But the letter arrived, and it confirmed everything she had heard, and as things turned out, this was the beginning of the end.

Two days later, on March 12, she wrote to B. J. Copeland, informing him that she had just been approached by the Greens. She had no idea how much toxin they would need or how often they would need it, but this information could substantially affect her allotment for toxin production, so before she could submit a revised budget, she needed clarification from him in writing.

For months now she had been looking for a way out of the funding process but had felt she needed a legitimate reason. Now she had just that, because there was no way that Copeland could respond to her request for clarification without acknowledging that he had not conducted a fair and competitive grant process. And if he refused to respond, she would be able to say that she was forced to turn down the funding because she could not submit a final budget when she heard that the director of Sea Grant, without her knowledge or approval, was promising free services from her lab to other researchers.

Many a time during these months, she wondered what had happened to the man she once esteemed as a mentor. B. J. Copeland was a legend among marine scientists in North Carolina. He had given many researchers in the state the support they'd needed to launch their careers, herself among them. Sure, he had his pet projects and his favorite researchers, which generated some resentments, but overall he had run a successful program at Sea Grant.

There were moments when she suspected that from the beginning he must have been in collusion with her foes at DEHNR and by helping to put her down he was firming up a relationship with a state agency that would continue to provide Sea Grant with money. She had no proof of this, but nothing else she could think of explained his handling of the process. There were other times, however, when it seemed to have been so sloppily managed that she thought maybe the arrogance that came with being a power broker for over two decades had finally caught up with him—that he probably had done things like this before but never been caught.

Though she did not know why he had turned on her, she realized now that she had been wrong about the B. J. Copeland she thought she knew, and that among his multiple motives were ones that would probably be forever unknowable to her. What she *did* know was that she felt thoroughly betrayed by him. And she would be wholly disgusted by the tactics he resorted to next.

He ignored her memo asking for clarification. Instead he began to put pressure on her. In public forums, in private meetings, during casual conversations in her absence, he spoke about how she was getting all this money to spend as she wanted and still it wasn't enough, omitting that he had directed where it was going. He branded her as uncooperative and blamed the slow progress with the *Pfiesteria* research on her unwillingness to share information and material with fellow researchers, never mentioning the fact that even without Sea

Grant moneys she had been supplying toxin to the coinvestigator on her proposal so he could proceed with the mammalian studies.

She repeated her request for clarification, and waited.

Then he got personal, suggesting that she was emotionally unstable, out of control, and ought to seek psychological counseling; that the inclusion of Mike Mallin on a component of her grant was inappropriate because they had a romantic relationship. He even brought up the loose-woman charge, accusing her of sleeping with Mallin and other students while they were still in her classes and implying that her lack of sexual morals spoke to her questionable character.

Repeating her request for clarification, she waited.

Finally, at wit's end, Copeland put an ultimatum to her: either she would submit a revised budget by July 22, so he could complete the funding award, or it would be withdrawn and redirected to Dr. Edward Noga.

Just before five o'clock on the afternoon of Monday, July 22, she had one of her students hand-deliver a letter to the Sea Grant office. It read:

"While all actions are subject to interpretation, and all interpretations must be regarded as subjective to some degree, I feel that the gulf which exists between my interpretation and your interpretation of events concerning the 'Pfiesteria grant process' is so vast that no easily achieved solution comes to mind. Accordingly, I sincerely believe that I must distance myself from this UNC Sea Grant funding process. The purpose of this memorandum is to inform you that, based on the requested clarification which was not forthcoming, and given the recent history of events, my colleagues and I shall not be submitting detailed revised budgets. Rather, we hereby withdraw our grant application from your program. . . ."

1 9

With her rejection of the money, what had until this time been a private struggle, conducted behind the scenes, surfaced as a public scandal that was covered by North Carolina newspapers. Not all of them took the same slant. With a front-page overview titled "Feud Stops Study of Fish-killing Algae," the *News & Observer* let everyone have his say, but gave B.J. Copeland an expansive opportunity to defend himself and, in the process, trivialize the dispute. Before this, Copeland had maintained that for the same reason the public should be spared witnessing how laws and sausage were made, the process of funding science should be conducted outside the purview of public scrutiny. Now, however, rejecting Burkholder's position that numerous irregularities raised questions about the fairness of the grant review process, Copeland said that he'd done nothing wrong, and the fame that had accompanied Burkholder's discovery had "mushroomed in her head." He attempted to shift the source of the conflict away from himself and toward the academic rivalry between her and Hans Paerl and the fact that she lost out on a key part of the study.

"There are all these egos clashing around here," he was quoted as saying. "I don't know what to do about it."

Paternalistically, he said he was trying to mend fences but was worried about Burkholder, whom he described as overworked and overly sympathetic to environmentalists who had a stake in her findings. The impression left by the piece was that the *"Pfiesteria* flap" was little more than a "scientists' spat."

Closer to the coast, at the *Sun Journal*, the daily published in New Bern, the reporting differences were striking. There, the journalistic lead was taken by senior writer Steve Jones, who, while researching a story about Burkholder's lack of cooperation with colleagues on the grant, had become bothered by the very issues that Burkholder had called into question. The thrust of his inquiry had changed after he examined each of the proposals submitted to Sea Grant, followed up with calls to the principal investigators, asked tough questions they had difficulty answering; he came away feeling that this story opened a door on a disturbing side of publicly funded science. In short, he did not think that scientists should be able to misrepresent themselves on proposals and then receive state funds, and he had begun to think that that was what had happened here.

As someone who took the spending of taxpayers' money personally —indeed, who thought of those who spent his tax money as his employees—Jones had taken his concern and findings to the highest level, requesting a telephone interview with the governor. During that conversation he pinned Hunt down, getting him to make a personal commitment to settling the dispute, and the result was a scoop for the *Sun Journal.* "Hunt Vows an End to River Study Ruckus," ran the front-page headline, as Steve Jones was the first to break the news that the governor was "not going to put up with any more of this fussing and squabbling." He was going to demand from B.J. Copeland "a full report on the funding process involving these public dollars."

After that, the editorial writers weighed in.

From the *Charlotte Observer:* "Gov. Jim Hunt ought to cut himself some verbal switches and take some bureaucrats to the woodshed for a good old Eastern N.C. thrashing. How come? Because from where we stand, it looks like bureaucratic bungling has gotten in the way of badly needed research on cleaning up our rivers. It's complicated, but we know this: Dr. JoAnn Burkholder is an ace N.C. State University researcher . . . and she's outspoken about state government's failure over the years to come to grips with riverine pollution. Her tendency to speak out has ticked off certain mid-level bureaucrats who don't appreciate her calling attention to their inadequacies. And many of Dr. Burkholder's supporters suspect that her bluntness may have been a factor when a series of state research grants were awarded not to Dr.

Burkholder, the world's leading expert on the problem, but to others who apparently are not as well equipped to do the job. . . . She was justified in telling the state recently to take its grant leftovers and stuff it."

From the *Sun Journal:* "The current shoving match over grant money for the study of a key piece of the Neuse River pollution puzzle should not be mistaken for another case of rivalry among scientists. Instead, evidence points to something vastly more serious—a calculated effort to put politics ahead of qualifications in awarding research work, with all the possibilities for poor science and waste of public money that could mean."

By this time public interest in the dispute was running high. Environmental groups, firm in their belief that Burkholder had run into trouble largely because she'd been willing to challenge the state's inaction and response to the pollution problem, petitioned the governor on her behalf. Private citizens were writing and calling in their support. Every day, it seemed, another article raised questions about the funding, and the one that ran on September 1 announced that a second top-level probe of the funding had been launched, this one by N.C. State's chancellor, Larry Monteith.

From the eye of the storm, JoAnn Burkholder watched and waited. The momentum of unfolding events allowed her to feel that after two years of delay and hardship, the outcome might actually favor her. It was by no means a certainty, and it was important, she felt, for pressure to remain on both state and university administrators. But it did almost seem that the shenanigans behind this story were destined to unravel.

When she went to bed on Thursday night, September 5, it appeared that the only thing that could stall things would be a disaster of greater proportions. Not that she was thinking in those terms. But then no one was. Like a sneak attack, the most destructive event in North Carolina history slammed into the state in the middle of the night, and caught everyone asleep.

~ ~ ~

Meteorologists had been tracking Hurricane Fran since it was a tropical storm in the Atlantic off the coast of East Africa, and warnings that she would come ashore around Wilmington had prompted coastal evacuations, but no one had any idea she would continue to gain strength, saving her final fury for Raleigh. The metropolitan area—sometimes referred to as a park with a city in it—was devastated by a record rainfall, which produced floodwaters and swirling winds that uprooted and toppled trees. One of the city's weekly papers would describe it as looking "like Bangladesh after a typhoon."

As did most of the populace, JoAnn Burkholder woke up on Friday morning and walked around her neighborhood in a daze, stunned by the devastation daylight made evident. Downed trees had snapped power lines, knocked down utility poles, and damaged transformers, which explained why there was no electricity. Her car radio informed her that more than a million homes and businesses were without power; it could be two weeks before some parts of the state received power again.

Her first thought was for the biohazard facility. Without electricity to run the aeration pumps, the cultures could crash and the ventilators would no longer maintain negative air pressure, so that toxic fumes would be released.

On a normal day, it took her fifteen minutes at most to get to the facility, but it took twice that on this day, because many streets were inaccessible, the signal lights at intersections weren't working, and traffic snarls slowed things to a crawl. Her fears were confirmed when she stopped at the trailer and found it without power, so she drove straight to her office on campus and began making calls. She tried to contact Howard Glasgow and was told he was on his way. Next she made a call to Physical Plant, the university maintenance department, and told them she needed emergency power. Nothing was available, she was told. Frantically punching more numbers on her phone, she tried to reach her dean, but he was out. Finally she got through to an assistant dean, and when she explained her dilemma he recommended she call Carolina Power and Light. Once she got through, she tried to explain her need; businesses with critical needs, such as major medical facilities, should expect to get power back first, but getting power to her research facility should be taken just as seriously. The best she got was a promise that someone would call her back.

"Okay, I will stay by this phone as long as it takes. But I need to know something. This isn't the same as butter melting. It is imperative that someone get on this."

When no one called, she phoned back to the assistant dean for advice, and he told her he didn't know what else to suggest short of calling 911.

If the stakes had not been so high, her conversations from this point on could be seen as almost comical. Calling 911, she was passed on to something called Emergency Operations. She identified herself as a researcher at N.C. State who was in charge of operating a biohazard facility where they were doing work with a toxic dinoflagellate, and they needed backup power because they had no electricity.

"I'm sorry," the voice on the other end of the line said. "A what?"

"Dinoflagellate."

"I've never heard of that. How do you spell it?"

Burkholder spelled it.

"What is it?"

To keep it simple, she said, "A toxic algae."

"I'm sorry. I don't know what you're talking about. Is it a bacteria of some sort?"

"No. It's an organism that emits a neurotoxin."

"Would you repeat that?"

With an exasperated sigh, she said, "An organism that releases a toxin that affects the neurosystem."

Still failing to get through to this person who had no idea of what she meant, finally hurling her pen in a display of absolute frustration, she all but shouted, "It's like nerve gas, for chrissake! And if it gets out, it won't be pretty."

From this point on, Howard Glasgow tells the story best.

After listening to Hurricane Fran howling outside his home in Durham all Thursday evening, he began receiving telephone calls around 12:30 A.M. As part of the modification of the dinoflagellate trailer to a Level 3 biohazard facility, he and Burkholder had wanted a backup generator, but lacking the money, they installed a security phone system, hooked up to sensors, that would call him with a recorded message if any of four emergency conditions existed. Alert Condition One was connected to a thermostat that let him know when the temperature inside the trailer went above or below the safe range. Alert Condition Two was activated if the exhaust fans went off. Alert Condition Three monitored the air pressure within the biochamber and the aeration system in the aquariums. And Alert Condition Four told him there was a sound problem: too much noise meant a motor had gone haywire.

At half past midnight, an Alert Condition Two call let him know that the power had gone out in the trailer. Glasgow punched in the code, verifying that he had received the call, waited five minutes, and called back, to check in case it was a temporary outage. He found the alert condition no longer existed, the fans had come back on. *Great,* he thought. Now he felt he could go to sleep.

Ten minutes later, the phone rang again, to report Alert Conditions One, Two, Three, and Four. *The roof must have blown off the damn thing,* he thought.

If that was the case, there wasn't anything he could do, so he punched in the code saying he had received the information. He was sitting there feeling helpless when he heard a crack outside, and a tree fell on his house. Narrowly missing the two cars parked in the driveway, it took out the front porch stoop. He didn't know until the next day that the house didn't suffer major structural damage. At the time, all he knew was that his house had a tree on top of it and there was something major wrong at the biotoxin facility. And preventing him from forgetting even briefly about the latter, there was a malfunction with the phone security system; he was unable to punch in the code that shut it off, and every twenty minutes he was getting calls reporting Alert Conditions One, Two, Three, and Four.

He didn't know what to do, so he let the phone ring every twenty minutes until around three-thirty, when he turned the ringer off on his phone and said to himself, *I'll handle it at first light.*

At six-thirty the next morning, he went outside to inspect the damage to his house. After sizing up the situation there, he turned his attention to the biotoxin facility and started making calls. He called the campus police and was told they could do nothing about it, he should call Carolina Power and Light. There, he got an answering machine and a pager, and he left his own number. All that morning, while dealing with the tree that leaned against his house, he continued to call CP&L. Only later would he learn that their answering machine was not functioning.

His immediate concern was getting aeration back into the tanks, because if he didn't, they would lose all the work of the four months in which the facility had been operating. They had only just begun to generate relatively pure toxin for the experiments the researchers on the *Pfiesteria* grant were planning to conduct, and without oxygen, the dinoflagellate would encyst, there would be bacterial contamination of the aquariums, and they would be back to square one.

Around two-thirty Friday afternoon, he left home; at three-fifteen, he arrived on campus. He had just walked into the basement lab when the phone rang. It was Burkholder, telling him that she had called 911, and a CP&L emergency response team was supposedly on its way to the biotoxin facility; he should go over and tell them what to do.

He didn't know who would be waiting for him, probably some line repairman, he thought, and as he made the turns that would take him to the trailer he relaxed slightly, believing that they would be back on line within the hour.

He headed up Sullivan Drive, and just as he crossed over Golden Street, he looked up the road and saw that a campus security officer had parked his vehicle at an angle, creating a roadblock.

Signaling that Glasgow turn around, the officer turned his attention to a line of cars that were pouring out of the parking lot of a housing area predominantly occupied by Asian students and their families. But Glasgow did not make a U-turn. Giving his horn a toot, he motioned the officer over and said, "I need to get in here."

"You can't come in," the officer replied. "There's been a hazardous-material spill, and we're evacuating the area."

"Well, I'm supposed to meet some emergency personnel from CP&L, and . . ."

At that point two and two began to come together, and he finished by saying, "Officer, I think I may be involved in this enough that I may be able to help resolve the problem."

After giving him a steady look, the officer pointed. "Pull in over there."

From that point on, things happened fast. A fireman with a hand-held radio came running up to see what the matter was. When Glasgow asked him about the spill, the man replied, "We don't know what it is." So Glasgow said, "You need to get me to somebody I can talk with. I think I can be of help." The man spoke into his cell phone, and within sixty seconds two men from the State Bureau of Investigation came jogging up. One of them grabbed Glasgow under his right arm, the other under his left, and they gave him a good hoist as they said "Let's go," and ran him one hundred yards down the road. As they came around a bend, Glasgow looked ahead and saw no fewer than five fire trucks; farther down, toward Western Boulevard, ambulances were lined up; and in the opposite direction, at least ten police cars flashed their blue lights. Meanwhile the evacuation of nearby buildings was proceeding, and he could hear men yelling, "Get the hell out. Get out. Get out."

Then he saw the hazmat team: fifty or sixty men completely encased in gold suits and helmets, forced-air respirators covering their faces. The SBI agents ran him right into the middle of this gathering, almost shoving him at the team chief, a man of perhaps sixty, who lowered his face mask enough to ask, "What do you know about this situation?"

Glasgow gasped out, "All I know is . . . they told me when I tried to enter here that there was a hazardous-material spill. . . . I came to meet a CP&L response team . . . to get power back at the Level 3 facility here. . . ."

Now the helicopters arrived. Two Blackhawks, with guns mounted on the front, hovered over them at a height of several hundred feet. Glasgow could barely hear the chief yell, "We received a terrorist call threatening the use of a nerve gas on Raleigh. I've ordered an evacuation of the area. . . ."

The rest was lost to the deafening beat of the propellers.

To make himself heard, Glasgow yelled back: "I think I know what you guys are talking about. We're growing an organism in a trailer down there that gases off a neurotoxin. But it's not a nerve gas. And it's not an immediate threat to the public. The only thing I need is to get power back on at the facility, and get the recirculating and filtration fans operational, and everything will be okay."

The chief knew what the protocol was, and he intended to follow it. He was also incredibly excited, and he did not know what he was dealing with. Shaking his head, he yelled, "That's not good enough. I want to know exactly what we're working with. I want you to give me the area of a hot zone in which we can work. Do I need to continue with my five-mile evacuation?"

"No, no, no."

"Do I need to send in these helicopters and incinerate this trailer?"

"No. Don't do that," Glasgow yelled, more concerned now about the hazmat team than he was about a backup generator. Because these people were *intense*. They had entered this situation thinking it was life threatening, and they were not interested in having him pat them on the back and tell them they were overreacting.

"Then tell me exactly what I need to do."

Thinking fast, Glasgow knew he couldn't tell the chief this was a false alarm. But at the same time, he knew he couldn't let the impression continue that this represented a threat to the city of Raleigh.

"I do not perceive this to be a public threat at this point. I perceive it as a threat completely contained within the facility. If I can take some people down there who can help me get the power back up, everything's going to be okay."

"You haven't given me a hot zone around the facility."

"The way the wind is blowing right now, I would say that in the worst-case scenario, if we had to worry about seepage of the toxin outside the facility, a ten-foot diameter would be sufficient."

The chief sent a look his way. "You hold on a minute," he said, and he backed off a way and called somebody on his phone. Glasgow assumed the chief was cutting back the extent of the evacuation.

Returning, he said, "If you can personally assure me that there is no immediate threat, and we can find some volunteers, I'll let you go down there."

At that, two people from Physical Plant stepped forward. "What do you need?"

"I need a power generator."

"We'll come up with something."

Glasgow looked back at the chief, who pointed at him and said, "This is your goddamn mess, you get it squared away. You go down there, you're on your own. I'm not sending my men down. But I will come down to inspect it myself once you get the situation under control."

The university personnel scurried off to find a generator. When they returned to the trailer with one that would work temporarily, Glasgow called the chief, and he came down for a look-see.

"Are you satisfied that there's enough electricity here to filter this room and that there is no longer a threat with accumulating toxins?" he asked.

"Yes, I'm satisfied about that," Glasgow replied.

"Okay. I have been in contact with officials at CP&L, and they have promised me that by sixteen hundred tomorrow, you will have full power restored to this facility."

"That'll be great. Thank you very much." And Glasgow shook the chief's hand.

In the aftermath of Hurricane Fran, the press would report on a variety of disasters. Pages would be given over to stories of homes and lives that were lost; about calamities caused by the loss of electrical service, of disabled sewage plants that discharged millions of gallons of untreated waste into the state's already polluted waterways. Everyone, it seemed, had a "Fran story." But missing from the coverage was the one about how the state's hazmat team saved Raleigh from terrorists who had threatened to unleash nerve gas on the city.

2 0

Throughout this period, the phone in JoAnn Burkholder's office rang
continually. During one stretch, her answering machine filled four
times in one day, then twice a day for two weeks after that. So
many people were complaining they couldn't get through to her that
she had the phone company expand the capacity of her voice mail
system from twenty to thirty calls.

Often it was members of the media calling, or university officials;
but she was also receiving calls from strangers across the country who
had read an article about her in a newspaper or magazine or seen her
on CNN or *Good Morning America*, wearing bioprotective gear as she
moved around her Level 3 laboratory, which rated on a par with those
for rabies and AIDS. They had been planning a vacation to North
Carolina or considering retirement there, and they wanted to know if
she thought it was a good idea.

She referred most of these callers to various publications, but some
calls she was willing to spend more time with. Calls such as those she
received from people who had come to North Carolina as tourists in
the summer of '95, whose children had gone swimming without realizing a fish kill was going on and had become sick afterward. She also
gave time to calls that came in from coastal citizens who were experiencing health problems they thought might be associated with *Pfiesteria* and were seeking her help because their family physicians were at
a loss. She even took one startling call from a pharmacist, who had
recently been inundated with people coming to her for over-the-

counter treatment of bleeding open sores and severe asthmatic problems and a range of symptoms associated with *Pfiesteria*. "Tell me what to do," she pleaded.

Burkholder wished she had more to offer them. If they had asked her what could happen if they went swimming in raw sewage and were exposed to E. coli or other bacteria, she could have given them a laundry list of different infections or diseases they might succumb to. But with *Pfiesteria*, none of that was known.

What was known was this: Mother Nature put mankind to shame when it came to designing chemicals that killed. Indeed, the very reason plants and animals formed toxins was to kill prey. But far less known were the mechanisms by which many of these natural poisons worked, because until there was a reason to study them—until, say, humans were adversely affected by them—no one cared. Then, once the study of a naturally occurring toxin was undertaken, the effort to identify and characterize the chemical(s) that made a poisonous species dangerous to humans was sometimes hit-or-miss, because researchers were starting with an unknown, and because in many cases a series of toxins, not just a single one, was released.

A case in point was the brevetoxin produced by *Gymnodinium breve*, the dinoflagellate responsible for the red tides in Florida and the Gulf of Mexico. Researchers had found that nine toxins, with different biochemical structures, were produced by a single cell. Not only that, but depending on its nutritional background—what the dino had been feeding on, the amount of sunlight it had recently been exposed to, and the current water temperature—the composition of those toxins changed. *Pfiesteria* was not a brevetoxin, but there was every reason to think that it too released a veritable army of toxins.

This complicated considerably the effort to decipher its effects, because multiple chemical compounds meant multiple means by which a toxin could attack a system. Using a stew as an analogy, a person could have an allergic reaction after eating it but wouldn't know what part he was allergic to: the celery? the carrots? the sour cream?

Multiple compounds also meant that different organs and physiological systems could be targeted in different ways at different rates. Certain chemicals in the toxin might cause an acute reaction, while others could work more stealthily, in a chronic or cumulative fashion.

A person's current state of health could even play into this, because some toxins had an unerring instinct for going after areas of inherent weakness in the host.

Another dimension that made this organism so formidable was that its ability to affect people was not confined to a single route of exposure. Unlike fish poisoning, where presumably you had to eat infected tissues to get sick, with *Pfiesteria* a toxic exposure was possible not only through ingestion but through dermal contact, entrance through cut skin, or simply by inhaling its aerosol. So there were four different routes of exposure, each of which had the potential to produce a huge variation in symptoms and in how severe they would be.

The toxin *Pfiesteria* released also appeared to be lipophilic: if the target survived the initial hit, it stored the chemical compounds within its fatty tissues. Depending on where these compounds ended up, how long they stayed in the body, and when they were slowly released back into the bloodstream, there could be a delayed onset of toxicity. Lipophilic chemicals could also be biotransferred to embryos or the milk of mammals, resulting in toxicity to offspring.

There was more bad news. Marine biotoxins were known to operate on what were called ion channels. Every cell in the human body has an ion channel, but certain cells, namely those whose function is to send a lot of signals elsewhere, such as the central nervous system and the immune system, are particularly sensitive to permanent channel disruption, unlike tissue such as skin, which after injury pretty much repairs itself and returns to normal. An injury to the immune or the nervous system leaves an echo forward, with the likely results being permanent immunological deficiency and neurobehavioral disorders.

With a host of other potential problems lying in wait—something that could cause cellular damage had the potential to create genetic aberrations, which in turn could disrupt the processes that governed cellular growth; read carcinoma—all of this explained why it was so difficult to come up with a clear set of clinical symptoms that could be blamed on *Pfiesteria*. The situation was far more complex than connecting, say, a parasitic microbe with its infected host. An as yet unknown number of toxins with four separate routes of exposure translated into an uncalculable amount of potential health outcomes, some acute, others subtle. A reduction in the body's ability to defend itself left it vulnerable to innumerable infectious diseases that might not have otherwise been threatening.

Most people, when they evaluated a health risk, thought in terms of damage that manifested itself in immediate outright disease. But chemicals need not operate directly and obviously in order to have dramatic consequences. People weren't used to thinking about exposures that caused insidious systemic damage or that bioaccumulated, creating long-term health problems. Or, for that matter, that interacted synergistically with all the other chemical insults people were exposed to in today's world. But this was all part of the *Pfiesteria* problem, which was why this organism posed such a critical public health challenge. Based on what was known, it could very well mean that instead of manifesting itself in an easily identifiable "signature" illness, the *Pfiesteria* syndrome was going to be an idiosyncratic pattern of health problems that masked the true cause or mimicked more well-known common diseases.

When a major health study recently conducted by the World Health Organization predicted that chronic conditions and new infectious diseases, not "dread events" such as AIDS and the Ebola virus, were going to be the scourge of the next quarter century, it did not cite *Pfiesteria* by name; but it was just the kind of thing its authors had in mind.

So it was difficult for JoAnn Burkholder to know what to say to people who called her to report they were suffering from a mysterious ailment after being out on the estuary, and asked: Could it be related to *Pfiesteria?* Almost anything was possible. But she couldn't say for sure. Until they knew more about the toxin, they were all in the dark.

One of her calls came from an osteopath in New Bern. A patient he was treating, a Neuse River fisherman and seafood dealer, was suffering from "the most severe case of cognitive impairment I've ever seen."

The osteopath went on to say that when he first examined David Jones, the man had so much trouble formulating a thought and verbalizing that the doctor concluded, *His cortex must be atrophied.* "It was unbelievable. It would sometimes take him ten minutes to get his thoughts together. And then all he could get out was a partial sentence."

After a series of tests ruled out Alzheimer's or gross neurological problems, he had put Mr. Jones on Dexedrine in an attempt to "increase the response between his synapses, and see if we could at least get him to where he could carry on a conversation." Being an osteopath, he had also planned to do some body work, because Jones complained of muscle aches in his lower extremities. Whenever the doctor laid his hands on a muscle group, however, it would spasm. "So I put him on Percodan for that, and at least he can get around a little better now."

"From what you could tell, is he otherwise healthy?" Burkholder asked.

"I did a physical on him, and he's in fine health. And he's not that old. Forty-nine."

She was quiet, thinking. "Well, can I ask why you are calling me?"

He told her he had been phoning around. "This is something I have never encountered before, so I wanted to see if I could figure out something to do about it." And after getting help nowhere—the state health department said they'd get back to him and never did—somebody at a local clinic had given him her name and said she might be able to shed some light on what was happening.

She was in no position to make a medical diagnosis, but from the sound of it, David Jones was living the hell that both she and Glasgow had gone through.

The offer came out spontaneously. "I don't know if it would help, but I'd be glad to talk with him. And with his family. Hearing from someone who has gone through something like this can sometimes make those to whom it's happening now feel better."

～ ～ ～

Route 17 from Raleigh to New Bern looked like Tornado Alley, there were so many uprooted trees and snapped power lines from Hurricane Fran. It rained most of the way, and what was normally a two-hour drive took her close to three hours to make. But the time passed quickly as she replayed in her mind her phone conversation with David Jones.

"How are you doing?" she had asked him.

In a halting voice he had replied, "Well, uh, I'm kind of getting along. I have good days and bad days."

"I know how it is," she'd said.

"I've . . . I've, uh, lost my train of thought, Dr. Burkholder. I . . . I can't think."

She'd winced, recognizing the fractured voice pattern.

"How's your family doing?"

"Well, now, they've been doing better the last couple weeks. . . . The doctor has been treating me. . . . And of course I've been . . . well, what I'm trying to say is . . . well, sometimes I don't know what I'm trying to say."

Toward the end of the conversation, Jones had all but broken down, admitting to her that he was frightened by what he was going through and afraid he'd never be normal again. When she offered to come and talk to him, he'd said, "Yes, ma'am, anything. Anything. Anything."

Jones lived in a bungalow just south of New Bern, on Neuse Drive, behind his place of business, the Neuse River Seafood Market, which had a *Closed* sign on it. He was a slim fellow, gentlemanly in demeanor, who looked like he'd spent a lot of time on the water: though he was not yet fifty, he could have passed for sixty. Accompanied by his wife and daughter, his movements slow and uncertain, he greeted Burkholder at the door and invited her into his living room, where for the next two hours he attempted to hold up his side of a coherent conversation. But it wasn't easy. Jones said he had taken his medication in anticipation of her visit, so he was feeling as good as he'd felt in days. But when Burkholder would ask a question, he would not be able to give her a complete answer without the help of his wife.

Burkholder tried to get him to put together a sequential account of his deterioration, and with great difficulty and much prodding, he did his best to cooperate. He said he'd noticed that the Neuse River was going downhill for at least ten years now—to which his wife nodded in agreement, saying, "I wouldn't let the kids go swimming in the river because my son got spinal meningitis and the doctor thought he got it from the river"—but it was only in the past few years that he'd begun to think that something was also going wrong with him. In his market, where he bought and sold fish and crabs, once there could be three, four people talking to him at the same time and he could answer every one of them and still be doing what he was doing, but increasingly he found himself getting confused talking to just one customer. About this same time, he found he didn't have the energy he used to have. Normally he'd be out of bed by five o'clock and he'd put a pot of coffee on and go to his desk and get things organized for the day, but it had become hard for him to get up, and he'd sometimes find himself at his desk listening to a buzz in his brain, and the next thing he knew, he'd fallen back asleep.

It had gone on like this for a year or so, he said, and then it had seemed to get worse. And if he had to say when that started, it was after the time he went fishing above the Neuse River bridge during a fish kill, and he was pulling in his nets and water splashed in his eyes. Immediately they started to burn, and then his nose started running and he was sick to his stomach. He could remember that incident clearly—"It's as vivid as I'm looking at you," was the way he put it—because right after that he got lost. "I mean confused, disoriented, didn't know where I was at." He compared it to being out on the water when a fog rolled in, which had happened many times; but whereas before he knew everything would work out because he could look at his compass and find his way back, or drop anchor and wait for the wind to pick up, this was entirely different. He didn't know what was happening. He wasn't sure everything was going to be fine. Nothing in his experience had prepared him for what he was feeling, which was panic.

In the succeeding months, his family noticed that he wasn't himself. He'd start to say something, and then it was as though his mind had wandered off. They would have friends over for dinner, and in the middle of the meal he would get up like he was going to the kitchen for something and not come back. He and his wife would watch a movie on TV together and he would be unable to follow the story line. They would make plans to go out for dinner, and she would remind him throughout the day not to forget, and he'd say, "Okay, okay," but when the time came to leave he'd get angry with her for not telling him they had plans, and by the time she lost her temper at him for forgetting, he'd be over it and wouldn't know what she was talking about.

By this time his wife was doing most of the talking, as she tried to communicate to Burkholder how disruptive the change in her husband had been to the household. His blowups were the most difficult things to handle, she said. For no good reason he would start screaming at her. "I know he doesn't mean to get mad," she said, "but it gets me all upset and I just don't want to be here no more." It was an explosive situation that she was afraid could turn dangerous, because some of the people he dealt with in his business were prone to violence.

She knew that whatever he was going through was hard on him, but it put a lot of strain on the people around him too, and it was especially hard on her. He wasn't functioning as a husband, he wasn't functioning as a parent, and he wasn't functioning as a businessman. When people called about business matters, instead of talking to him they began to ask for her.

And then that went under. He had closed the doors on his seafood market when they temporarily banned fishing on the Neuse the previous fall, and when the ban was lifted several weeks later, he did not reopen. He couldn't concentrate on the paperwork. He refused to answer the phone because he was unable to carry on a conversation. He would get up and have a cup of coffee and go out and get behind the wheel of his pickup truck, and he'd sit there letting it idle for sometimes an hour, trying to remember what he was supposed to do, before he turned the ignition off and trudged slowly back to the house.

At that point, neither of them had any idea what the source of his problems were, and the first doctor he'd gone to wasn't any help. He had acted as though Jones was making things up. But after the examination, when Jones walked out of the office, the same fog that had settled around him on the river descended again, and for almost an hour he had wandered around the parking lot in the rain, looking for his car.

So now they were a family in distress. He wasn't working, because he couldn't work. On his "good" days he would call a friend and they would go out on the water and fish. And he liked to mow the grass, even though, as his wife interjected, he didn't mow grass the way anybody else did. It was over here, over there, and before he was done with his own yard he'd be off mowing the neighbor's. Meanwhile the household had been turned upside down. She'd had to take over the responsibility of making all the important decisions; and because someone had to generate an income, she had gone to work full time.

"I'm the breadwinner in this family now," she said, making no attempt to disguise her bitterness. "And my friends don't understand, my mother and his mother don't know what's going on, and they all look at me like, well, they think he's going through a nervous break-down or something. And I try to explain, and they just say, 'Well, leave him. Give it up.' I know it's not his fault, but I don't know how to deal with it anymore. I don't know how much longer I can live with a person like this, even though I love him."

As this woeful tale unfolded before her, Burkholder kept assuring David Jones that he was not alone in what he was suffering from and trying to find the words that would console his wife. But Burkholder was also very specific in what she thought he needed to do to turn this thing around, because she knew that a basic principle of toxicology was that if you were going to have any chance of recovery, you had to terminate ongoing exposure.

"If this is from *Pfiesteria*, you have to stay off the the water, David. No more fishing or crabbing."

When Jones made a face, she said, "I know. It's probably the last thing you want to hear. But that's the only thing that we know works. The slightest exposure to *Pfiesteria* at this point is too much."

Fishing was all he knew, he said. It was what he'd done all his life. If he couldn't work and couldn't fish, what was he good for? he asked her.

Feeling his pain deeply, Burkholder shook her head and said, "All I can tell you is that as far as we know, it's the only thing that's worked." Then, softening her voice, speaking as someone who had been there and come back to tell about it, she said to Jones, "And the thing is, David, they miss you. Your family misses the person you really are. They want you back, and that would be the only way."

As Burkholder drove away from New Bern late that afternoon, thoughts and images and snatches of conversation crowded her mind. Sitting and watching David Jones struggle to express himself and re-membering what it had been like for her had been an uncomfortable experience, made all the more difficult by her having to bear witness to the devastation his illness had wrought on his marriage. It made her angry all over again at the health department, which she believed might have been able to prevent something like this from happening if it had just acted responsibly in the first place. And should anyone there claim it had, she was prepared to offer as evidence two letters Jones had shown her.

Just before she got up to go, she had asked Jones if he'd had any dealings with the state. Nodding, he said that he had received a call from a health official who was doing an epidemiological survey. Jones had done his best to answer the man's questions, but it had been a struggle to carry on a normal conversation, so an appointment had been made for him to be examined by a neuropsychiatrist from Duke University Medical Center. The results were delivered by mail, and he brought the letter out for Burkholder to see.

Noting that he had severe problems with "attention, memory and learning," the examiner had concluded that Jones's foremost problem was depression. Without attempting to address its source, the doctor had then written a prescription, and in the space for diagnosis he had written *Pfiesteriosis*.

But it was not necessarily an official admission that *Pfiesteria* was the source of his troubles, as Jones discovered when he wrote the state health director, Dr. Ronald Levine, to see if there was any way the state would be willing to provide health care or financial assistance to someone who had been afflicted while working on one of the state's waterways. The response had come from Dr. Michael Moser, director of the Division of Epidemiology: "Scientific research has not established a cause-and-effect relationship between exposure to *Pfiesteria*-related fish kills and any adverse human health effects. While such an association has been alleged, the data available to support the allegation do not meet generally recognized scientific standards of proof. Some people who report that they were exposed to *Pfiesteria*-related fish kills have also reported symptoms, but this does not establish causation. Staff of the Division of Epidemiology are not aware of the presence of any established toxin in the Neuse River at levels sufficient to cause human illness."

Once, when Burkholder and B.J. Copeland were still talking, she had asked him what she had done to fall out of his favor.

He'd laughed heartily and said, "Has anyone ever called you paranoid?"

Failing to see the humor, she had remarked, "I may be paranoid, B.J. But even paranoid people can have enemies."

Each time another piece of information or personal testimony came to her attention that, in her mind, contributed to the idea that the North Carolina Department of Health had engaged in a conspiracy against her and was deliberately attempting to suppress information about *Pfiesteria,* JoAnn Burkholder would recall that conversation. She knew it was a serious charge to make and one she could not prove with hard facts; but she also knew what had happened and did not believe for a moment that it was an accident. To borrow from Rick Dove's analogy, when the streets are wet, it probably rained.

According to those who have been in a position to observe the health department in North Carolina over time, it has a history of extreme conservatism when it comes to ascertaining environmental risks, especially when the evidence suggests the source might be industrial or agricultural pollutants. Avoiding public scrutiny about how it internally evaluates the seriousness of a health threat other than to say it needs a definitive scientific statement before it takes action, the department traditionally has proceeded with such caution in instances where direct epidemiological investigations were called for that in a number of instances it did nothing until faced with federal intervention, or until pressed by concerned citizens or the media. At which time it finally acknowledged the need for an epidemiologic study, and then one of two things happened: it took its time, emphasizing the difficulties of the undertaking; or it rushed the results, more often than not concluding that its investigations produced no unusual findings.

In defense of this "crisis avoidance" approach, the department has invoked the public interest, saying that if a perceived threat has not been documented, it is important not to create undue public alarm, which could spark widespread panic, negatively impact commercial interests, or bring about the imposition of costly environmental regulations. Understandable and worthy reasons all—unless they become overriding considerations, which is the dark side of the department's conservatism.

In the opinion of one former health official, a prime directive within the health department is "noise control." Wherever possible, staff is encouraged to figure out ways to keep a low profile, unless there is an opportunity to show that the department is doing a good job of protecting citizens. If something comes up that makes it look as if the department isn't doing its job, "you do everything you can not to make the newspapers, or to be discussed in the governor's office or the state legislature. Because then you run the risk of upsetting people who have control over your budget or say over whether your job is secure."

It was no secret that Dr. Ron Levine had survived multiple administrations and fifteen years as state health director and assistant secretary of health—a notoriously vulnerable job—because he knew how the game was played.

In a model state health department, if information had been brought forward by a reputable scientist about a potent environmental pathogen that was killing another species by the millions, and included were reliable reports that strongly suggested it was a health menace to humans, a protocol would have kicked in. Toxicological data banks would have been checked to see what other toxins might compare to this one. A laboratory determination of the toxicological properties of this organism would have been initiated. Exposed people believed to have suffered illnesses would have been interviewed in detail. Simultaneously, an epidemiological inquiry would have been mounted, along with a "disease prevalence" survey at hospital emergency rooms and local health clinics, and among physicians in the areas where people affected would go for treatment. And all this would be followed by a vigorous, comprehensive environmental and health study of guaranteed integrity.

A version of this protocol was pursued by the North Carolina health department, but a point-by-point look at its implementation raises serious questions about how truly interested the department was in the right answers. Or whether it was engaged, instead, in its customary effort of minimizing public concern.

■ The whole emerging disease and marine biotoxin issue is something new for most public health agencies. There are in the environment a variety of substances hazardous to people's health that public health officials do not understand and are unprepared to deal with. They don't test for them, and therefore they don't regulate them. And when one of these comes along, officials typically try to explain it away with what is known rather than delve into what it might be. It is an inherent flaw in the system.

But at the time Burkholder first approached the health department, there was a large body of knowledge available about paralytic shellfish poisoning, neurological shellfish poisoning, and, since 1987, amnesiac shellfish poisoning. It was clearly understood that some algal toxins, such as those associated with the red tide, could become aerosolized, because people going to the beach during the red tide of 1987 had complained of respiratory problems. So even though there was very little data, outside of the work done by Burkholder and her colleagues, to support the notion that an estuarine dinoflagellate might be aerosolizing a neurotoxin that was harmful to humans, a straightforward theoretical connection existed, based on the literature alone.

Combined with anecdotal reports, the appropriate public health response would have been to err on the side of caution and aggressively launch an inquiry as if there were in fact a problem. To say: If there is something here, how do we find it? This is a very different approach from allowing institutional caution to throw a shadow of doubt over JoAnn Burkholder's reports and taking the position that *Pfiesteria* was innocent until proved guilty.

In interviews, Levine denied that he and the department were looking for reasons to discount what she was saying; but at the same time, he was unable to account for why, at the very start, he allowed the environmental division of DEHNR, which was engaged in a public dispute with Burkholder, to undermine so easily her credibility. When Dr. Levine was told that Burkholder had not published anything about *Pfiesteria* in a peer-reviewed journal, his source had confused the American magazine *Nature*, put out by the Smithsonian Institution for the general reader, with the prestigious British scientific journal *Nature*. What was surprising was not just that Levine didn't know better, but that he had been willing to take Steve Tedder's word over Burkholder's on such an important issue.

■ Outside scientists I consulted about the next point rendered a unanimous opinion: A scientific understanding of the toxin produced by *Pfiesteria*—its biochemical makeup, how it was created, and how it concentrated in a system—should have topped the order of business. Once that box was opened, you could determine what biological damage was being done and what clinical symptoms to look for. You could begin to develop biomarkers—biological measurements that were indicators of exposure—and even prevention measures and detoxifying compounds.

A toxicologist within the health department had argued as much at an early meeting of the various technical professionals on staff. "The toxin should be characterized before anything else," Dr. Kenneth Rudo told the gathering. "If we want to do a risk assessment, if we want to do a meaningful epidemiological study, we have to know what we're talking about from the standpoint of what's causing the problem. We know about its life cycle, but it's still vague in terms of what it is and what it can do to humans. You have to have that before you know what to look for and can understand what you're seeing."

He had gone on to say, "If you like—and this is an offer—I can shop it around intramurally at EPA [the Environmental Protection Agency] or NIEHS [National Institute of Environmental Health Sciences]. I think I can find someone to do it and have an answer within a year. And it won't cost us a dime."

For reasons never given, Dr. Rudo's offer was not picked up on. And because this basic first step was not taken, over two and a half years later, the toxin has yet to be characterized.

■ There is an ongoing battle in the environmental hazards world over the funding by industry of academic institutions to conduct research that leads to equivocation about proof that a given industrial chemical is harmful to humans. A somewhat similar controversy exists in the public health field, where some departments, after considering the political and economic implications of a positive finding and anxious to convey a reassuring message to the public, will proceed with a premature or flawed epidemiological study that all but guarantees an inconclusive finding. This enables the department to say that it has looked into the possibility of a health hazard and at this time there is not enough evidence to establish a health outcome. In this case, the epidemiological inquiry initiated by the North Carolina health department was so plagued with bias, uncertainty, and methodological weakness that it not only was inherently incapable of accurately discerning the extent of the risk posed by *Pfiesteria,* but left a legacy that stands as an impediment to future undertakings.

Dr. Peter Morris of the Occupational and Environmental Epidemiology Section took the epidemiological lead for the department, and it began with four phone calls on a Sunday afternoon to fishermen whose names had been supplied by Burkholder. The men he spoke with told him they were suffering from a variety of health problems—neurological, respiratory, intestinal—but unable to discern a "nice, tight pattern" that would allow him to define the illness he was going to study, Morris concluded that "something was amiss." He interpreted the absence of a clear set of consistent common symptoms as a warning sign that Burkholder may have been exaggerating, and he reported to Dr. Levine his suspicion that there might be nothing to this.

Later Dr. Morris would defend his "exploratory investigation" as a preliminary first step and say those who found his tone and questions offensive were misperceiving him, but he would have more difficulty justifying the two epidemiology studies that flowed from it. About the study that was promised to a group of researchers from East Carolina University, he would beg off, saying his "superiors" told him to contact those people and refer all other questions to Dr. Levine. About an in-house epidemiological study that the health department itself would go on to conduct *after he had reported his doubts that anything healthwise was going on,* he would say that it was a departmental decision that, in retrospect, was a "big mistake."

This "mistake" was underwritten by $15,000 taken from the $600,000 designated for *Pfiesteria* research. It consisted of a more extensive phone survey, asking people who thought they might have been exposed to a fish kill if they had developed any significant symptoms, and a rapid-response study. When a fish kill occurred, Morris and a team of health officials were supposed to go down to the coast and question and examine people within days of their exposure to look for a consistent pattern. But "mistake" does not adequately account for two large holes at the very center of the study.

For one thing, according to Morris, he was relying on staff at the Department of Environmental Management's regional office to inform him when a fish kill was taking place and put him in touch with people who had been exposed. But when asked about the arrangement, the DEM people he was counting on said their role had been peripheral at best, because they never took the study seriously. They had been given no guidance or resources to cover their involvement. Management had not convened a meeting with health officials to discuss how to coordinate the effort. As one DEM biologist said, "We saw it as posturing by the health department, so if there was a crisis from *Pfiesteria* their asses were covered."

For another thing, without a scientific understanding of the toxin, the compounds that contributed to its toxicity, and a clinically established set of symptoms that were linked to exposure, it would be impossible to establish a cause-and-effect relationship between *Pfiesteria* and any illness.

So what was the point?

"It would let us know if there was evidence of *major* adverse health effects," Dr. John Freeman, the now retired director of the Occupational and Environmental Epidemiology Section, would state when questioned about the value of the study. "I mean, it's good information to be able to say, 'Hey, we don't expect you to die if you keep fishing.' "

Maybe. But what it most assuredly would not provide was any useful information about the sublethal damage caused by *Pfiesteria*. It didn't provide any insight into whether *Pfiesteria* was causing reduced immune capability, for which there was a battery of tests available, or nervous system malfunction.

What it did was enable the health department to say: Based on the available evidence, there is no discernible association. . . . And while the study told nothing about whether *Pfiesteria* was a public health threat, it became the baseline inquiry that any subsequent study would have to refute before it would be accepted. In short, it introduced yet another confounding factor into the search for truth.

■ On numerous public occasions, when called upon to explain the department's skepticism toward *Pfiesteria*-caused health effects, spokesmen would say that if people were being hurt, that information would have turned up when the department contacted hospitals and physicians on the coast. However, a quite different story was told by numerous members of the New Bern medical community when they were independently approached for confirmation of this claim.

"We don't know whether or not there's a health risk," said one physician, who echoed the sentiments of others. And he explained why. "When it comes to external infections, we culture for the standard things you get if you cut yourself cleaning up your garage. We're not going to recognize a wound caused by some unique waterborne organism. The local lab isn't going to culture for it either. We would simply prescribe a broad spectrum of antibiotics."

The same went for many internal conditions—liver disorders, viral illnesses, high fevers: "We treat many of them without ever knowing the causative agent. And if the patient gets better, we don't pursue it."

As for hearing from state health officials about *Pfiesteria,* a physician who has been in practice in New Bern for eighteen years, first in emergency medicine and now as an internist, said no one ever contacted him or, as far as he knew, any of his colleagues. So removed from reality did he feel public health officials were when it came to what local doctors were seeing in their patients, "There would have to be a catastrophic event—people would have to be dropping dead on the beaches—before the health department would react."

Still another said, "If there's a problem here, and I'm sure there is, the state doesn't want to know about it. In fact, they're terrified something might turn up, and they don't want to have to address it. They would rather turn a blind eye to it and hope the problem goes away. But trust me, it ain't gonna go away."

■ This certainly wasn't the impression given by Ronald Levine when he wrote B. J. Copeland on June 2, 1994, to notify him that the department intended to transfer over a half-million dollars to Sea Grant to fund a *Pfiesteria* study. "This . . . is a matter of serious concern for our Department," he said, using the words "urgent" and "vital" to characterize what appeared to be research the department desperately needed so it could fulfill its responsibility to protect public health.

But if Dr. Levine genuinely felt a sense of urgency at the time he wrote the letter, a manifest lack of follow-up indicates that it didn't last very long. The normal period of time between an identified research need and an RFP is six months. In this case, nine and a half months would pass before a Memorandum of Understanding between the health department and Sea Grant would be signed, and it would be a simple two-page memo, notable for its lack of technical details and its excess of generalities. Then another five months would go by before an actual Request For Proposal was mailed, with the announcement of results coming in late November, a year and a half after the original letter of intent was sent out.

If the *Pfiesteria* research was so important to the health department, what took so long?

In the course of several long interviews, I repeatedly pressed Dr. Levine for an answer. At first, he said, "I don't know." He admitted that it was an inordinate amount of time and the department had been slow in moving, but he denied there was any "horsing around" on his part. Asked a second time, he replied that he had been frustrated by the length of time it was taking and had tried on numerous occasions to call B.J. Copeland and get an explanation for the delays. When he finally did get through, he said, Copeland blamed it on "the bureaucracy." Before asking a third time, I spoke with Copeland, who denied having any such conversation with Dr. Levine or that he was a difficult person to get hold of. So the next time I spoke to Dr. Levine, I played a tape recording of Copeland's denial and, one more time, asked, "What took so long?"

After a silence so protracted I wondered if he'd heard my question, he said, "No comment."

"No comment?"

"No. You heard what I said, you heard what he said, you figure out who's telling the truth."

I had one more question, and I led into it this way.

"Dr. Levine, I want to be clear about something. Throughout the entire year that you are waiting for an RFP to take place with this money, you are still committed to the research. Is that right?"

"Absolutely."

"Okay. Then would you please explain what you were thinking when you wrote this E-mail, dated August 23, 1995, a year and two months after your letter of intent to Sea Grant?"

I handed him a copy of the memo. It doesn't matter how it came into my hands; what matters is that on that date, Dr. Levine wrote to Steve Levitas, deputy secretary of DEHNR: "We have set aside almost $600,000 of 'health' reversions to be funneled through Sea Grant for studies of dinoflagellates associated with fish kills. 18 months have gone by without spending a dime of this money, to my knowledge. Since malfunctioning septic tanks in Craven County may well be an inciting factor in diminished water quality in the Neuse, I suggest we lop $25,000 off that fund before the money actually reaches B.J. Copeland for distribution to contractors. If you . . . agree . . . I need to take immediate action with our budget office."

The memo concludes with a parenthetical note in which Dr. Levine states that if it could be pulled off, he would be willing to "take another $75,000 from this fund to beef up our training of sanitarians."

"It was just something to be considered," Dr. Levine said. "Yes, I put it on the table. But we ultimately decided not to do it." And, he assured me, "It wasn't because I didn't think the *Pfiesteria* project was important."

I nodded, he looked as if he didn't feel like talking anymore, and shortly afterward I ended the interview. I'd gotten my answer.

■ There are some very competent and conscientious people working for the health department, who do a fine job of protecting public health, and who feel that public scrutiny of what transpires internally is a healthy process. Within that group are some who are profoundly disturbed when the department as a whole is tarred by the poor performance and negative approach of a few. It is from these people that I learned the tensions between Dr. Greg Smith and JoAnn Burkholder erupted into a vendetta.

After being upstaged by her during the public meeting in New Bern, for which he was reprimanded by DEHNR secretary Howes and instructed to take a training course in public sensitivity, Dr. Smith, I was told, was furious with Burkholder's attitude. Seeking and receiving the approval of Section Chief Dr. Freeman, they adopted an adversarial approach toward Burkholder that took the form of an unwritten policy that the department would no longer make an effort to work with her.

There were those who disagreed with this course of action. They felt it was inappropriate to cut off, just because she could be difficult, a researcher who was the primary source of information on a potential health risk. It was to be expected, they pointed out, that researchers and health practitioners outside the state bureaucracy would be passionate about their findings or feel their first allegiance was to their patients, and the department was wrong to react defensively when there was a conflict.

Over those objections, the department ceased all cooperation with Burkholder and even went on the offensive. One tactic, pursued by Dr. Smith, was to make unreasonable demands for data she possessed relating to *Pfiesteria*, so he could accuse her of being uncooperative if she was not forthcoming. Another was to lobby receptive members of the scientific panel awarding *Pfiesteria* funds, in an effort to make sure that as little money as possible flowed her way.

Longtime observers of the health department watched the "*Pfiesteria* flap" unfold with an intense sense of déjà vu, because they felt it resembled a previous case the department had botched so effectively that it could have been a rehearsal for this one.

Throughout the 1980s, Caldwell Systems Inc. operated a commercial hazardous-waste incinerator in Lenoir, a small town in the western part of North Carolina. Over the years, a growing number of people living close to the incinerator and working for CSI began to suffer from respiratory ailments and neurological complications. Support for the idea that the incinerator was making people sick came from a state air-quality inspector by the name of Roy Gorman, who reported that there were no animals, no bugs, no snakes, no birds, in the vicinity of the plant, and documented that the emissions were damaging vegetation. A local physician, Marc Guerra, diagnosed dozens of residents and CSI workers as having a variety of illnesses—running sores, tremors, short-term memory loss, behavioral changes, and autonomic dysfunction (sweating and swings in blood pressure, temperature, and heart rate)—which he attributed to exposure to toxic chemical waste.

"It was clear to an idiot what was happening," Dr. Guerra would say later, but despite the overwhelming evidence, the state environmental officials responsible for regulating the plant said it was "in substantial compliance" and refused to close it down. Gorman was told, "There is no law against killing trees in the state of North Carolina," and for speaking out of line was transferred to a dead-end desk job. Dr. Guerra, meanwhile, was branded as an alarmist.

Under extreme public pressure from a coalition of Lenoir residents and environmental groups who accused the state of putting the operation of a hazardous-waste incinerator above public health, State Health Director Ronald Levine agreed to authorize a study to determine whether there was a possible "causal relationship" between incinerator emissions and the ailments. And he appointed Greg Smith to lead the investigation.

Within the health department and without, opinion on Dr. Smith was divided. No one questioned his intelligence. A Phi Beta Kappa graduate from the University of North Carolina–Chapel Hill, with an M.D. from the UNC School of Medicine and a master's in public health from the same institution, Smith had all the right scientific and medical credentials and affiliations to be an outstanding figure in the public health field. He was bright, at times brilliant. But he had a reputation for being cocky and combative, for enjoying center stage in controversies. It was also said that Smith had upon occasion made hasty judgments before he'd done his homework, and rather than going in with an open mind and seeking facts objectively, he tended to pursue data that buttressed his position. It was also said that Dr. Smith had a tendency to give offending industries the benefit of the doubt.

Because Smith was the chief author of the report, people expected that its conclusions would back up state regulators and protect the company. So when it came out, no one in Lenoir was surprised to read that he had been unable to discern a distinct health syndrome, nor was there enough evidence to confirm or rule out the possibility that pollution from the plant harmed people living nearby.

It was slipshod work, and it would come back to haunt the health department. There was by this time so much public distrust of the state and so much negative publicity that the EPA took charge. Physicians for the federal Agency for Toxic Substances and Disease Registry conducted their own investigation, and their findings invalidated the state's conclusions. They found not only that the rash of illnesses was caused by exposure to toxic chemicals, but that contamination levels merited a site cleanup that qualified the incinerator for the nation's Superfund list of most hazardous sites.

In the aftermath, Dr. Levine would apologize to the community for not acting quickly and for allowing the problem to get worse. He would offer the excuse that the department had not been prepared to undertake a health-risk analysis. Acknowledging that the state's "inaction" had put citizens at risk, he would assure everyone that the department had learned from its mistake. And he would say, "What we need to be discussing now is not recrimination or finger-pointing but healing, learning and preventing such a combination of circumstances from occurring in the future."

The way *Pfiesteria* was handled would allow the department's critics to cite historical precedents, maintaining that the pattern it represents has become clear with pathological enormity.

Indeed, the parallels are striking. There was an initial skepticism expressed about the health risk posed both by *Pfiesteria*'s toxin and by the incinerator's emission, which made both residents and local physicians feel the department wasn't truly interested in finding a problem. Not coincidentally, Dr. Smith was the department's point man, and just as he publicly accused JoAnn Burkholder of turning *Pfiesteria* into hysteria, in public forums in Lenoir he criticized Marc Guerra, the local physician who relayed residents' and workers' complaints, for his advocacy. In both cases a study was undertaken without an understanding of the offending agent—at CSI there was no waste-stream analysis of the incinerator's emissions, just as *Pfiesteria*'s toxin was not characterized—and the report issued was so inconclusive that people would claim it was intended to produce ambiguous results. And in the end, just as the newspaper in Lenoir would report, "From Day One, it was a cover-up, runaround, whatever was necessary," a similar accusation would be made regarding *Pfiesteria*.

It also should come as no surprise that within the department, JoAnn Burkholder was disparagingly referred to as "Roy Gorman in skirts."

For almost two years, officials on the administrative level at North Carolina State University had given different signals when it came to supporting JoAnn Burkholder. Sending out one set of signals were university officials who had been taken aside at meetings by representatives of agribusiness and reminded that N.C. State was a public institution that depended on public support, the implication being that Dr. Burkholder was a liability. Very different signals had been sent by officials who, while they would have preferred to avoid controversy, were aware there had been a shift in priorities within the university. Times had changed. Where once they had been concerned only with excellence and quality in their faculty, to that had been added relevance, regardless of the consequences, to the needs of the state. And they knew that if you were going to be relevant, you ran the risk of getting involved in controversial issues.

Joining the latter group were those who knew that JoAnn Burkholder was the biggest news story ever to have come out of the academic arm of N.C. State. National and state media; the scientific press; radio and TV—the woman was in demand. The closest they'd seen to it had been the two-or-three-day burst of publicity that flared up around the N.C. State biostatistician who testified on DNA forensics at the O. J. Simpson trial. The university wasn't keyed into the benefits of this kind of publicity—couldn't put a dollar figure on it or say that students were flocking to the university to work in Burkholder's lab—

but there was no doubt about it, recognition for her was recognition for the campus.

None of this, however, mattered to Vice Chancellor Charles Moreland, the man who had been given the task of determining whether misconduct had taken place once the money for *Pfiesteria* research had been transferred to Sea Grant, which administratively was part of the N.C. State system. A former chemistry professor in his mid-fifties with thinning brown hair, a sleepy demeanor, and black-framed glasses so heavy it looked as if they had lowered the bridge of his nose, Moreland had had limited experience in matters like this: investigating a few charges of plagiarism and falsification of data. But about Burkholder's outspokenness, he had no problem. Having been a professor, he had strong feelings when it came to the issue of academic freedom. If a research faculty member did good science that could be backed up with facts, he felt she had the right to speak out, pro or con, even though it made some people uncomfortable. And as for the responsibility of investigating the Burkholder-Copeland dispute, he accepted it with a mind that if anything improper had gone on, he intended to get to the bottom of it, because to do anything less would be to share complicity.

But as he rapidly learned, saying you were going to get to the bottom of something was easier than getting there. After a meeting with Burkholder, at which she detailed her contention that an "unethical research 'give-away'" had been guided by the granting-agency director, who had also disseminated vicious rumors about her, Moreland was incensed. If what she said was true, then as far as he was concerned B.J. Copeland deserved not only censure but termination. But Moreland wanted to hear both sides, all the arguments; and when Copeland set out his version of the events, he made it sound as if the process had been carried out fairly and professionally. His explanations for the irregularities that so disturbed Burkholder appeared plausible. It made sense to broaden the research to include multiple investigators, he said. You wanted the most efficient and effective mix of research. There were no done deals before the RFP was conducted. Discussions had been conducted with prospective researchers, but that back-and-forth wasn't a guarantee; they still had to go through a review process. As for misrepresentation on the proposals, what was important was not what you claimed you could do but what you accomplished, and on that you should be judged. About giving money away after the process was officially over, discretionary allocations by granting agency directors were not uncommon. And concerning slanderous remarks, it never happened. Burkholder was just bitter because she thought she should have got more funding.

Going beyond he said/she said and sorting out the truth drew Dr. Moreland into a spiraling quest that took over a month to complete, when he had expected to settle the matter in a week or two at most. He was a man who had the intuitive ability to zero in on the heart of a problem, but in this case he felt an obligation to substantiate his conclusions incontrovertibly, because he knew that in the end he would not be the one making a decision. His report would go to the chancellor, Larry Monteith, who in turn would make a presentation to the president of the University of North Carolina system, C. D. Spangler, Jr., and he would decide on any subsequent action.

By the end of October he felt he had sufficiently scrutinized B.J. Copeland's actions and answers—even allowing for multiple takes on the same event—to arrive at a judgment. While he was unable to verify that the Sea Grant director had improperly manipulated the RFP process, or to confirm that the whole thing had been plotted between DEHNR and Copeland, Moreland had uncovered a disturbing number of inconsistencies that appeared to contradict Copeland's "official" version.

Perhaps most glaring—and certainly most symbolic—was Copeland's declaration that the reason Hans Paerl had been given the nutritional ecology component of the grant was that the reviews on Burkholder's proposal were not as good. After examining all the proposals and all the reviews, Moreland found the exact opposite to be the case. From what he could determine, Copeland appeared to have manufactured the idea that Burkholder had received negative reviews. In fact, she had been given a rating of "excellent" in every category of every review. One reviewer had gone so far as to write: "of all the proposals in this series that I have read, this one stands head and shoulders above the rest and must be funded."

The question became: Was it fair to assume that the best proposal should get funded? Burkholder received the best reviews, yet someone else got the money to do the work that she was recognized as the expert on. When she was given the money, it was essentially to do the lab work for others.

Moreland never did touch bottom. His investigation turned up "a lot of talk, but not a lot of hard information." And in part, that was precisely what justified his conclusions. The written documentation—the paper trail that would either have exonerated Copeland or convicted him—was nonexistent. It was impossible to say whether the proper science protocols and procedures had been followed. The records that were kept gave such a sloppy and partial picture of the funding process that its integrity was left open to question. And doubt. There was only B.J. Copeland's word to say it had been done fairly, and that wasn't good enough for Moreland.

Nor was Copeland's denial that he had made derogatory statements of a personal nature about Burkholder, for Moreland was able to corroborate them from multiple independent sources.

There was no letter of explanation. Since Copeland served in an "at will" position within the university, he could be replaced without cause. Further, the university did not want to open itself to legal action. B.J. Copeland received a phone call from a university official requesting a seven o'clock meeting at a Bojangles restaurant on the outskirts of Raleigh. He showed up without any idea of what was coming, and he was told, "You're out. By December 31."

<hr />

When Jim Hunt said he was going to meet with the feuding academics and resolve this flap, it was obvious to his aides that nobody had briefed the governor on its history. They understood why he had said what he had. A sharp journalist had put him on the spot, and he had responded in typical Jim Hunt straightforward style: By golly, I'll just call them into my office and we'll figure out what the problem is and we'll fix it. But the aides did not think it was a good idea. They knew it wasn't going to be as easy as he thought, and this was an election year. So the role of dispute mediator was tactfully removed from the governor's plate and passed to Jonathan Howes of DEHNR.

Like a lot of people who have enjoyed political power for quite some time—he had been mayor of Chapel Hill—Howes was used to a press that accepted easy answers and went away. He was also used to settling in-house conflicts by treating them as misunderstandings that could be cleared up simply by giving people a chance to explain themselves. So it was not surprising that his "investigation" consisted primarily of writing a letter to B.J. Copeland, conveying the governor's concern, and asking him for his account of the *Pfiesteria* funding process.

In reply, Copeland provided Howes with a notebook tracking the sequence of events and the accompanying correspondence, assured him that nothing was wrong with the funding process or with the science used in the studies, and expressed his hope that "we can put this matter behind us and proceed with the work needed to improve coastal water quality."

Just to be on the safe side, Howes then asked the counsel for DEHNR, Richard Whisnant, to look into this matter and see if anything aroused his concern. And after conducting several conversations with DEHNR officials and reviewing the records and finding nothing in violation of the two-page Memorandum of Understanding with Sea Grant, Whisnant reported that he had turned up no wrongdoing. With this, Secretary Howes concluded that as far as DEHNR was concerned, maybe things could have been handled better, but he was satisfied nothing in the broad processes had violated the terms of the memorandum, Sea Grant appeared to have fulfilled its obligations to establish a focused inquiry into the matter, and the best thing now was to move ahead and make sure things like this didn't happen again.

The governor's internal inquiry would probably have ended there had it not been for Marion Smith, the deputy director of Hunt's Eastern North Carolina office. A petite woman in her forties, with enormous brown eyes and matching hair, which she wore in a bun, Ms. Smith had a deceptive Southern sweetness that masked political canniness. She was the kind of appointee who reflected positively on the governor's ability to surround himself with quality people, and as events would transpire, she would prove not only her loyalty to her boss but her worth.

While Howes seemed to be satisfied with B.J. Copeland's account, she was not. She was cognizant of the governor's campaign promise the previous year that a priority of his administration was going to be to clean up the Neuse River, and that symbolic of his commitment to water quality had been his pledge to help JoAnn Burkholder with her *Pfiesteria* research. She also knew that the governor was aware that Burkholder had been critical of what the state had done to protect water quality, and that much of her criticism was pretty much on target, which was the reason he was supportive. So she decided to look into this issue on her own to make sure that Hunt could not be embarrassed by any unexpected developments at a later date.

There was another reason behind her actions. She had been present, two years earlier, at the meeting where Levine told Burkholder he was going to let Sea Grant administer the funds. And it had been her distinct impression that Sea Grant was supposed to put together a group of disinterested, appropriate scientists to serve in an advisory capacity to Burkholder, to review her protocols and procedures and to resolve the credibility issue, but that unquestionably Burkholder would be the one conducting the research, because she was the expert. As Marion Smith remembered it, there had been no mention of a competitive grant process.

When she reviewed her notes and correspondence and first began to talk with people, Ms. Smith did not really expect to turn up evidence of a conspiracy within state government to downsize Burkholder. But when she read the Memorandum of Understanding, she was appalled by its vagueness and brevity. She had expected a thick, detailed document, not two pages, most of it boilerplate, with only six lines that defined the project criterion. It was hard for her to understand why, if this was such an urgent situation that it had to be funded by reallocating reserve funds, it took almost nine months from the time the money was identified to get a two-page memorandum and final transfer accomplished.

Then, when she learned that the researcher who had been awarded the epidemiology component wasn't an epidemiologist at all but an anthropologist, and that the reviews of his proposal were so bad some members of the committee had questioned whether it should be funded at all, "I nearly fell over."

But the clincher came when she contacted Dr. Patricia Tester to find out why she had withdrawn from sitting on the scientific review panel. Dr. Tester said that after she was asked by Copeland to serve, she had seen a newspaper article and heard a discussion on National Public Radio that indicated some of the funding had been prepromised. She had phoned B.J. Copeland to get clarification, but he never returned her call. Then she had faxed a copy of the article, with a quote from Will Rogers on the cover sheet—"I know only what I read in the newspaper"—and asked, "Can you confirm?" Again she received no response. So she had asked to be excused, making up a conflict-of-interest story.

Marion Smith had heard all she needed to hear about B.J. Copeland. She had always held scientists in the same high regard she did priests and nuns, as people dedicated to the search for truth without bias, without prejudice. Her job being to make sure government was responsible to citizens, she knew that taxpayers, who supported most of the scientific research that went on in this country, wanted answers to questions such as: Does a specific environmental problem cause disease? Adding significance to this dispute for her was the fact that this was not just about getting the science straight; this was a potential public health crisis. She felt the citizens of the state had a right to expect the best solutions, and she knew that honesty was critical to this process.

About DEHNR's role in all this, and whether there had been an orchestrated effort to shut Burkholder out of the research and conduct a "wired" funding process, she was reserving judgment until she learned more. But try as she did, she got no closer to an answer. For a crime to have taken place, there had to be motive and opportunity, and both were present. She would come to believe that higher officials in DEHNR *had* to have known what was happening and that there was no way it could have been done if B.J. Copeland and DEHNR weren't in it together. But whether that meant there had been a meeting at which everyone sat around a table and came up with a premeditated plan, or there were winks and a concurrence that this was the way it would be played—the way most back-room politics were conducted—she just couldn't say.

Unable to connect the dots in a way that would spell "foul play" for everyone to see, Marion Smith did feel that, at the very least, DEHNR owed JoAnn Burkholder an apology. To some, that may have seemed like a hollow gesture, but having grown up in a society that in general tended to value women much less than men, and aware of how much more of a struggle it was for women to receive recognition than for their male counterparts, Marion Smith knew that public apologies mattered to women.

No one authorized her to write a letter, but she knew that if she waited for someone at DEHNR to get around to it, it wasn't going to happen. So she wrote a first draft and delivered it personally to Secretary Howes, saying that since the university had taken action against B.J. Copeland, it was time for closure on the part of DEHNR, and here was the form it should take.

Howes didn't have a problem with that, but Dr. Levine did. When the draft was circulated for comment, he quibbled over the language in the apology, because he did not want it to appear that the department was "heaping praise" on Burkholder. He thought it was unnecessary to refer to Burkholder as a "fine researcher," that calling her a "well-regarded researcher" would do just as well, and he fought for that point until eventually he got his way.

On November 1, 1996, Secretary Howes sent Burkholder a letter saying he regretted the past problems between her and DEHNR and assuring her they would not happen again. He ended by expressing the desire to "forge a new relationship that will allow us to move forward on our common goal—improving water quality in North Carolina."

The tweaking of the language in the letter had not surprised Marion Smith. It was almost to be expected. The important thing was that the Department of Environment, Health and Natural Resources had publicly admitted it had made a mistake. And if the whole story were to be told, greater gratification had come from a remark by someone who knew the history of this whole sorry situation, who said, "It's such a shame that all the balls in this administration are to be found under skirts."

~~ ~~ ~~

It would be nice to be able to say that on the wind of these two developments JoAnn Burkholder was finally vindicated, her ordeal was over, and she sailed into the future as the symbol of the heroic, dedicated scientist who took a stand against the system and prevailed. But when reporters who called for comment tried to bring up those lights on her saga, she would have none of it. She wasn't even comfortable when people expressed admiration for the way she had adhered to her convictions. True, she had put her career on the line in a showdown over principle and honor, which was no small thing. But the perception she wanted people to have of her was simply that she was an honest scientist who felt morally compelled to bring her research and its implications forward. It just so happened to be groundbreaking research in a controversial area.

In her mind, there was nothing "heroic" or especially admirable about that. The way she saw it, "I don't understand why more people wouldn't think that way. Why they don't have respect for what's around them. Why they don't think it's a sin to play games with people's health. I don't get their way of looking at life."

When an article in *Natural History* put her in the same company as Rachel Carson, author of *Silent Spring*, who revealed that cancer-inducing chemicals remained as residue in virtually everything we ate or drank, she rejected the comparison other than to say they both "saw troubling things going on."

As anxious to put this "mess" behind her as she was to move forward, she turned her attention to writing a new proposal. The money she had rejected, about $250,000—everything that had been originally awarded to her except the money that had gone directly to the researcher doing the small-mammal studies—had been transferred from Sea Grant to a fund within the university, which was going to control its distribution. "Write what you need, beyond the confines of the remaining allotment if it's appropriate," Dr. Moreland had told her, and she did. The thrust of the new proposal took her beyond technician status. She would still be producing toxin so it could be characterized and its metabolic effects on mammals understood, but she would also once again be running a research lab. And in addition to the funds that would allow her to analyze more precisely the nutrient recipe that *Pfiesteria* loved, she intended to also ask for improved equipment: a decent microscope; a more permanent facility that would permit them to work with more than one culture at a time and wouldn't blow away in a hurricane; and, of course, a backup generator.

She wasn't sure this dream list would be fulfilled, but she was told by Marion Smith that when Governor Hunt was informed that there might not be enough of the reallocated money to underwrite what needed to be done, he had said he would find a way to put it in next year's state budget.

While she welcomed this news, she was cautious about getting excited, because she was still wary about the way the system worked. She knew, for instance, the background to the "official" apology from DEHNR and the arm twisting that had taken place before it was written. She believed she remained detested by higher officials in the agency for challenging them in a direct and public way, and that they would do it all over again if they thought they could get away with it.

Fueling this suspicion was a public debate that was being played out on the pages of the New Bern *Sun Journal,* which was providing a forum for State Representative John Nichols from Craven County, who was tenaciously pursuing the idea that there had been serious misrepresentations on two of the *Pfiesteria* grant proposals. Nichols had been bothered by the fact that Dr. Hans Paerl stated in his proposal that a fellow researcher would be able to identify *Pfiesteria,* when in fact that researcher could do no such thing. He also didn't like East Carolina University's team having written that to connect illnesses suffered by fishermen with *Pfiesteria* outbreaks, they would use maps provided by a researcher who did not possess them—for only Burkholder did. He wanted an accounting for those discrepancies, because from where he stood, it looked as if someone was lying.

And since taxpayer dollars were funding these proposals, Nichols had written to the governor, asking for an explanation, only to receive what he felt was a smooth dodge: that neither Jonathan Howes nor the attorney for DEHNR, Richard Whisnant, in investigating improprieties, had considered the accuracy of the information contained in the grant proposals and whether the researchers were making claims they could not support. Nichols's next letter was to the state auditor, requesting that he "investigate this apparent fraud." And in mid-December a spokesman for the state auditor said they had determined there was sufficient evidence to warrant an investigation—the third into the funding process.

Although she had her opinions, of course—"If a scientist can say whatever he wants on a proposal and excuse it by saying that as long as he produces results its okay, he's saying the end justifies the means, which is eyebrow-raising"—Burkholder kept her distance. She was glad, however, to see large questions being asked about the way science should relate to the public trust.

And indeed those discussions were now taking place. It was a topic for guest writers on the editorial pages: "Because public funding for environmental management research is intended to support decision making, I contend that scientists proposing such research have an obligation to give public officials—in addition to good science—some idea of how likely it is they will produce information useful for the intended purpose." And it was talked about in the government halls in Raleigh: "If an academic institution wants to pursue research that improves the field of knowledge, that's all well and good. But in terms of the funding we do through state government, if we're doing research on coastal environmental concerns, the taxpayers have a right to expect that their money is being spent for something that's really worthwhile. And that's the lesson for us. We need to better manage the system of how we administer these kinds of grants."

It goes without saying that none of this talk mattered to *Pfiesteria*. As JoAnn Burkholder had said all along, "*Pfiesteria* hopes the state continues with business as usual. It's doing just fine, thank you. And with increasing water-quality degradation, it will do better."

That was part of what made the two-year delay so unfortunate. There was more *Pfiesteria* out there now. Its population bank was larger. Not only that, given the boat traffic and other opportunities for transport, it was likely that it had been transplanted to waters else-where—along the coastal United States and even to coastlines world-wide.

The other unfortunate aspect was the health price people continued to pay, because if Burkholder was right, then not only had people been hurt; they were still being hurt.

Until the fall of 1996, the evidence to support that contention, though compelling, was circumstantial. But that was before Dr. Ed Levin, her coinvestigator, completed his first series of neurobehavioral tests, in which he attempted to model human exposure through the use of laboratory rats.

The studies were conducted in his laboratory at Duke Medical Center. He used albino rats, which were a benchmark species, very well characterized in terms of their learning mechanisms. He put them in a radial-arm maze, which was like a wagon wheel without a rim. And taking advantage of rats' natural tendency to explore new places and their sweet tooth for the cereal Froot Loops, he conditioned them to seek-and-find and enjoy a reward; then he injected them with aquarium water laced with relatively high doses of *Pfiesteria* and assessed their working memory, or "blackboard memory" as it was sometimes called when the objective was to determine whether an exposure was able to erase what had been imprinted.

Levin was aware of the importance of his findings, so he repeated the experiments several times, then ran them by his peers for comment. Even when he was able to replicate the results and his colleagues could find no flaws in his methodology, he wanted to be extra cautious about making extrapolations to human beings, or about getting ahead of his findings and "starting the dominoes to falling." Because the results were alarming. In his tempered terms, the "magnitude of deficit" was "highly significant." The injected rats lost their "choice accuracy" by 50 percent, which was enough to make anybody sit up and pay attention.

And officials of the EPA did just that when word of the results reached them. They asked Levin to come to the offices of their main research laboratory, in Research Triangle Park, and present his results.

Twenty-seven people attended, which was a lot for a 4 P.M. Friday meeting. And after Levin brought everyone up to date with his latest experiments, there was sufficient apprehension among the toxicological experts present that they were not interested in working with the organism in their facility, but enough excitement was generated that they asked him to let them know how much money he needed to take his research to the next level.

JoAnn Burkholder was at that meeting, and as she listened to Levin, her thoughts wandered. Levin's studies were the evidentiary hole card linking *Pfiesteria* to cognitive impairment in mammals. For three years she had had to put up with skepticism and disbelief. As recently as that summer, someone in the health department had been quoted as saying to a German television crew that both she and Glasgow had fabricated their experiences in order to raise funds for their research. She was sick and tired of it, and tremendously relieved now that another scientist—

one who was credible and came from an excellent institution—had corroborated, through controlled experiments, what she had been alleging all along.

Almost as important to her was that, unlike so many other people throughout the sad history of this research, Levin hadn't tried to pat her on the head and say, "We'll take it from here, little girl. Just give us the toxin." They were colleagues in the research, and that respect meant a great deal.

By the same token, it gave her an uneasy feeling to think that this was just the first step in understanding the full range of adverse effects the toxin had on humans, of which short-term memory loss might be the least.

Pfiesteria was a harbinger of bad news: that had been the message she'd preached from the very beginning. But whereas, when she first said it, she had been commenting on water quality in North Carolina and how pollution was changing nature, her message now took on multiple meanings. It spoke to the fact that pathogens of this kind were on the rise—*twelve years ago there were twenty-two species of known toxic dinoflagellates, now there were fifty-nine*—and that with more pressure being brought to bear on estuaries and coastal waters— *75 percent of all Americans lived along the ocean and Great Lakes, and the coastal states were growing fastest*—environmental conditions were shifting in their favor—*in the summer of 1996, Pfiesteria was blamed for the death of twenty thousand hybrid striped bass in a Chesapeake Bay aquaculture facility.* Already, *Pfiesteria* had been found in waters from Delaware Bay to coastal Florida to the Gulf of Mexico, and was the prime suspect in many unresolved kills elsewhere. And she did not doubt but that there were other, similar species lurking in the sediment around the globe, that they had been for millions of years, and that other disasters might already be happening. They just weren't expressing themselves spectacularly but were wreaking silent havoc and going undetected because of our limited ability to trace the link between microbial toxins and physiological effects. And because local governments usually tried to cover the connections.

We've all known that if we continue to put population and pollution pressures on the environment, sooner or later limits will be reached and the earth will strike back. The questions are: What will that look like? Will we recognize it when it happens? How will we respond?

Until now, we have depended on other species to be our early-warning system. But this defense only works if we keep our eyes open to what is happening around us. One could argue that with deformed frogs turning up in the wetlands of Minnesota with withered arms, stumps for legs, and eyes in their throats; fish rolling over in the estuaries of North Carolina; and now these human accounts, the line between us and them has been crossed. But doubters have always been able to counter that the case is weak because the evidentiary bridge between species has not been built.

That was before a 1996 study was completed into the sudden and mysterious die-offs of manatees—marine mammals also known as sea cows—in south Florida in record numbers. Autopsies by marine scientists identified pneumonia as the cause of death, but there was no explanation for why the manatees were suddenly so vulnerable to that particular illness. There did appear to be a time-lock association with a red-tide outbreak—the same red tide that drifted up to North Carolina in 1987—but for a long time scientists were unable to figure out whether or not there was a connection; if there was, what the killing mechanism was of the toxin, *Gymnodinium breve;* and why adult manatees were dying but juveniles weren't.

Then last fall they developed a fluorescent marker that enabled them to tag the toxin. And when they put the lung tissues of a dead manatee under a microscope and added the marker molecules that would react with the toxin to make it glow, the samples lit up as though plugged into a light switch. Next, using another set of techniques, scientists were able to determine that for some reason the toxin collected in the lungs of manatees, and over a period of time, as it accumulated, it had programmed the cells to age prematurely, weakening them until finally the lungs were unable to resist infection, allowing bacteria to come in and do the rest.

Although the dinoflagellate involved in red tides is not *Pfiesteria piscicida,* what this new understanding of manatee mortality demonstrates is that marine mammals chronically exposed to toxic dinoflagellates can harbor the toxin in vital organs until a threshold is reached, which explains how sublethal effects can lead to death. It also speaks —clearly, compellingly, and analogously—to the human implications.

It was too early to estimate the human toll taken by *Pfiesteria,* but sometimes Burkholder found herself wondering: *What if it had been worse? What if this had been a relatively small outbreak of an infectious disease that had the potential to grow into a global epidemic if it wasn't quickly checked?*

E P I L O G U E

"I ain't never seen a dead man spend no money."

It was a typical September morning in eastern North Carolina. Seen from the coast, the sunrise was an open fire, but just a few miles inland, where a thick mist hovered over the tobacco farms and kudzu-wrapped tenant shacks, it was a flame barely visible through an old smoky kerosene lamp.

In order to make my 8 A.M. meeting, I had got an early start, but a school bus was setting the pace, which was a little slow for me but allowed me time to romance the sights. Otherwise I might not have noticed the overnight harvest of roadkill—fresh coon and opossum—littering the narrow, shoulderless road that cut a corridor through thick piney woods; the small white country churches with sharp spires set like candles to light the spiritual way in clearings that looked chopped out of a wilderness; two black women wearing bonnets and fishing with bamboo poles and earthworms for yellow-bellies and white perch, from a bridge that spanned a quiet creek the color of sweet tea. In a country diner where I stopped for coffee, I overheard a farmer and a fisherman arguing good-naturedly about who had it roughest, and could tell that neither of them would be willing to trade places and find out.

Two days earlier, JoAnn Burkholder had heard another horror story from a stranger on the phone. The caller this time was a wholesale seafood dealer who bought fish, shrimp, and crabs straight off the boats that worked Pamlico Sound, and he was sending an SOS. The "boys," as he referred to the fishermen he did business with, were scared to death of what was happening on the water, he'd said. He had followed her plight in the newspapers, seen that officialdom was trying to shoot the messenger, and he was calling to let her know "there's a lot of people who can back you up on this stuff. A lot of people."

She wanted to go down and meet with him in person, but there was a scheduling conflict—she had to fly to D.C. to speak at a national conference on habitat loss in estuaries—so she got his permission to let me come in her place.

My experience with fishermen and crabbers over the previous year had been so full of paradoxes that I had no idea what to expect. You would think that of all people, they would be among the most environmentally enlightened of folk, because they were the ones who saw firsthand that when human endeavors did bad things to the environment, there was a rebound effect, causing bad things to happen to humans. Don't get me wrong, some were. Dallas Ormond was. But there were a surprising number who didn't take the time to read or attend meetings, so they weren't informed about what was happening and were content to say, "We've got people who are supposed to take care of that," and leave it at that.

I remember going out on a boat one afternoon with a commercial fisherman who was going to show me a "dead zone," where nothing lived and where the last time he'd gone overboard he'd come out "eat up" with sores. The sun was blazing, a light chop ruffled a bay once used by pirates, and we were skimming along the surface, talking, when he downed the last of a bottle of Red Dog and tossed it into the wake. He could tell by my disapproving glance what I was thinking— *Aren't we talking about mending our pollution ways?*—and cupping his ear, he shouted, "Do what?"

I shook my head, first in answer to his question, and again when he cheerfully rationalized that he was creating, not ruining, habitats by providing shelter to baby crabs. "Think of it as an artificial reef."

Not long after that exchange, he seemed to cross into a different realm, and when he cut the motor and drifted over a sea grass bed that had once been productive for him and said, "Here's where," his tone was sober.

From the deck of a bobbing boat there was nothing unusual to see, just a dark shade of green that said the bottom here wasn't sandy. But using a rake with a hollow metal handle that amplified the nick when it dragged against the hard surface of a shell, he hauled in a half-dozen clams, and all were dead. I asked for a turn and got the same results. I was going to ask him to tell me again who he thought the killer was, just to hear the words he used, but when I looked at him I found him gazing over the water as if it could still hurt him, and decided nothing else needed to be said. I did note, however, that when he finished off his next Red Dog, he put the empty back in the cooler.

The dealer who had telephoned Burkholder did not want the fact that he was calling to be made public, so his name and any identifying features of his seafood house are omitted. Suffice it to say that shortly after eight I found myself sitting in a cinder-block office that smelled of sea wind, talking to a wholesale distributor who ships seafood caught in North Carolina waters to markets throughout the United States. And the story he told put this deplorable situation into perspective.

He said he had about a hundred fishermen and crabbers who brought their catch into his house, so he was in a prime position to observe the strange happenings that had the "boys" spooked. Before this, he said, about the worst that could happen was you'd fall overboard and drown. But now people were dying mysteriously: from infections that started out as scratches, not gashes; from conditions that shouldn't have been fatal; one from a liver disease the doctor said must have been cirrhosis, except the fellow didn't drink.

"There's a lot of people that worry about that water out there," he said. "I mean, there's a bunch of them." And he went on to talk about "places on the water where if you run through it you start throwing up. That happened to me twice this year. And I know another boy it happened to." He'd also heard from the crew of a forty-five-foot shrimp trawler that passed through an area this summer and everyone was knocked out. They actually collapsed on the deck and got up woozy.

It was hard to spot these zones until you were already in them, because "You never see nothing, you never smell nothing." The only thing that distinguished them from anywhere else was that if you fished the area, "you won't catch nothing. Whatever it is that's there, it kills everything." Everything in the crab pots was dead. "It even kills oyster fish, and hell, they're hard to kill with a hammer."

Conscious of my speech—the night before, a motel desk clerk, unable to place my outsider accent, had asked me if I was from New Zealand—I asked if this information had been brought to the attention of the appropriate authorities.

"We did. After them boys on the trawler got up, when they got in here we called the Department of Environmental Management about it. And when they come by, we thought we were in trouble."

"What did they say?"

"They said we were just trying to stir things up, there wasn't nothing wrong with the water, and they were gonna put me out of business if I kept up about it. That's how they do you. They say if you keep on about how bad things are, you're only going to make it hard on yourself, 'cause all they'll do is come in and post your area."

That was the reason, he said, that whenever the state did a study on health effects, "they don't find nothing wrong. They can't get nobody to say nothing. The boys are almost as scared of being put out of work as they are of what's on the water. They got a family to support. They can't take a chance on losing their livelihood."

At the same time, he said, they believe, as he does, "that it gets to a point where something's got to be done about it, and nothing will unless someone says something."

It was like he told a dealer from Albemarle Sound who reported they were seeing some of the same things up there, but he wasn't going to say anything because he was afraid the state would close him down: "I told him, 'Hey, I ain't never seen a dead man spend no money.'"

We still don't know whether *Pfiesteria* will be a plague upon our waters of Old Testament proportions. Crucial scientific questions remain to be answered. But one thing is certain. Even when its mysteries are solved, it won't be over.

Glossary

algae Primitive plants that may photosynthesize as higher plants do but mostly lack vascular tissue (and therefore have no flowers, roots, stems, or leaves). Dinoflagellates are generally considered to be algae, mostly because the most infamous species, the red-tide formers, are capable of photosynthesis as higher plants are. Zoologists, however, consider dinoflagellates to be animals (protozoans).

algal bloom Rapid algal growth that usually results in a discoloration of the water.

anthropogenic (source of nutrients) Derived from humans (sewage) or human activities (e.g., crop farming, high-intensity animal operations, automobile exhaust pipes, urban runoff).

chlorophyll The photosynthetic pigment responsible for converting light energy to chemical energy used for plant growth.

cyst A protective structure produced by some algae as part of their life cycle or in response to adverse conditions. The cyst has a thick protective outer wall, and the cell within may contain a high quantity of stored food reserves.

dinoflagellate Solitary organism, usually one-celled, and with both plant and animal affinities. The correct pronunciation for *dino* is with a short "i," from the Greek word for "whirling," but the term is often pronounced with a long "i," indicating the Latin word for "terrible" or "terrifying."

estuary An inlet of sea reaching into a river valley as far as the upper limit of tidal rise.

eutrophic High in nutrients (nitrogen, phosphorus) and high in organic (biological) production.

eutrophication The excessive addition of nutrients, which spurs accelerated algal growth, creating more plant biomass than the ecosystem is capable of using.

neurotoxin A toxic substance that interferes with the functioning of the neurological system. Dinoflagellate neurotoxins typically block transmission of nerve impulses, so that death occurs from muscle paralysis and resulting suffocation.

nitrogen A biologically important nutrient essential to plant growth, which exists in solid, gaseous, and liquid states. Nitrogen supply regulates plant growth in North Carolina's estuarine waters.

nutrient Usually, elemental substances (nitrogen and phosphorus) used by plants to grow.

phosphorus A mineral nutrient also required for growth, which exists mainly as phosphate, a dissolved solid.

photosynthesis The conversion of light energy to chemical energy. Plants use water, carbon dioxide, and sunlight to manufacture sugars that are used for growth.

phytoplankton Microscopic, photosynthetic plants that are suspended in the water column.

pigments Large, colored molecules that capture light energy and make it available for photosynthesis.

plankton Organisms, both plants and animals, that are suspended in the water column and transported by tides and current.

Acknowledgments

In the sense that this book could not have been written without the help of many people who gave generously of their time and their trust, this is a collaborative work. For personal and professional reasons, some have expressed the preference not to be thanked by name. Two people I would like to single out for special appreciation are Goethe "Bud" Aldridge, who introduced me to this story and served throughout as a research assistant, and Dr. JoAnn Burkholder. It wasn't always easy for her to open herself and give what I asked for, but in the end we are all, I believe, richer for it. Two other people I want to thank are referred to in this book as Dr. Peter Cover and Janice Kishiyama, because they requested that I not use their real names.

From the North Carolina Department of Environment, Health and Natural Resources, I would like to acknowledge Barry Adams, Jim Mulligan, Kevin Miller, Steve Levitas, Jonathan Howes, Debbie Crane, Steve Tedder, Linda Rimer, Jim Overton, Dr. Ronald Levine, Dr. Greg Smith, Dr. Peter Morris, Dr. Kenneth Rudo, Dr. John Freeman, Dr. Michael Moser, and others.

From North Carolina State University, I would like to acknowledge Dr. Charles Moreland, Dr. Johnny Wynne, Dr. Eric Davies, Howard Glasgow, Tim Lucas, Dr. Edward Noga, Michael Dykstra, Bruce Macdonald, David Rainer, Dr. Larry Monteith, Dr. Gerald LeBlanc, Dr. Durwood Bateman, Jim Easley, Dr. B.J. Copeland, and others.

From the state of North Carolina, I would like to express my appreciation to Derb Carter, Randy Waite, Dr. Ed Levin, Dr. Donald Schmechel, Cecil Hobbs, Dr. John Costlow, Mike Mallin, Aaron Schecter, David Griffith, Dr. Hans Paerl, James Pinckney, Bob Lucas, Bill Lotz, Dr. Marc Guerra, Roy Gorman, Stuart Leavenworth, Brett Childers, Dick Trammel, Aileen Glasgow, David Moreau, Kristin Rowles, Phil Shaw, Dallas Ormond, Jim Pounds, Etles

Henry, Jr., James Guthrie, Dean Ahrenholz, Katy and Glenn Blackburn, and others.

From the New Bern area, I would like to express gratitude to Marion Smith, Rick Dove, Phil Bowie, David Jones, George Wetherington, Steve Jones, and others.

From places elsewhere, thanks go to Dr. Stephen Smith, Dr. Paul Epstein, Dr. Richard Clapp, Dr. Paul Auerbach, Dr. Brian LaPointe, Michael Straight, Catherine Werner, Terri Gilmore, and Janet Bailey.

From my publisher, I want to acknowledge Eric Rayman, Toni Rachiele, Rebecca Head, Carol Bowie, and Pamela Duevel.

Finally, I would like to mention my agent, Anne Sibbald, of Janklow/Nesbit Associates, who helped shape the proposal for this book; my brother Brad Barker, who functioned as my media watchdog; and my wife/photographer/proofreader, Star York.

Index

Levine and, 188–93
life direction of, 23
Mallin and, 61–63, 185
on Marine Fisheries Commission,
175–79, 212
media coverage of, 68, 70, 72, 85–
86, 174, 288
nature revered by, 179, 199
new grant proposal of, 318
NMFS's site visit to lab of, 125–
126
Noga and, 32–35, 39–40, 46, 68–
73, 82–83, 87–88
North Carolina State University
and, 310–11
office of, 52
Ormond and, 170–73
Paerl as professional rival of, 270–
271
paper of, published by *Nature,*
83–85
parents of, 198–99
Pfiesteria grant proposal of, 268–
271, 272–73, 274, 275–77
Pfiesteria-induced illness of, 90–
94, 95–96, 107–8
Pfiesteria life cycle described by,
43–45, 77
phycology course of, 36, 37
and pork producers, 262
presentation of, at Toxic Marine
Phytoplankton Conference, 65–
67
public notoriety of, 86
public support of funding for,
280
recommendations of, to APES,
163–64
research funding and, 60–61, 73–
74, 97

Sea Grant and, 226–27
at Seventh International
Conference on Toxic Marine
Phytoplankton (1993), 106–10
social life of, 26–27
as speaker, 174–75
Steidinger and, 101–3
swine lagoon rupture documented
by, 236–39
Tedder's animosity toward, 214
and timing of *Pfiesteria*
emergence, 232–33
toxic dinoflagellate culture
transported by Glasgow and,
77–78
toxic dinoflagellates tracked by,
45–47, 57
and victims of *Pfiesteria*-induced
illness, 288–89, 291–97
at Water Quality Subcommittee
meetings, 181–83
writing ability of, 83
business vs. public health, in North
Carolina, 259, 262

Caldwell Systems, Inc., 307–9
Cape Fear River, poultry lagoon
rupture on, 247
Carolina Power and Light, 281, 283–
284, 286
Carson, Rachel, 318
Castaneda, Carlos, 151
Centers for Disease Control (CDC),
13, 95, 125, 252
Charlotte Observer, 280–81
Chesapeake Bay, 66
children, *Pfiesteria* illness
susceptibility of, 11–12, 15,
252, 257
Chowan River, 220, 221

About the Author

Rodney Barker has been a newspaper editor, an investigative reporter, and a feature writer for a variety of regional and national magazines. His previously published books include *Dancing with the Devil: Sex, Espionage, and the U.S. Marines* (1996); *The Broken Circle* (1992); and *The Hiroshima Maidens* (1985). He lives in New Mexico.